T0136237

Canals and Communities

Arizona Studies in Human Ecology

Editor
Robert McC. Netting

Associate Editors
Peggy F. Barlett (Emory University)
James F. Eder (Arizona State University)
Benjamin S. Orlove (University of California, Davis)

Edited by Jonathan B. Mabry

CANALS and COMMUNITIES

Small-Scale
Irrigation
Systems

University of Arizona Press   Tucson

The University of Arizona Press
© 1996 Arizona Board of Regents
All rights reserved

⊛ This book is printed on acid-free, archival-quality paper.
Manufactured in the United States of America

01 00 99 98 97 96 6 5 4 3 2 1

Library of Congress Cataloging-in-Publication Data
Canals and communities : small-scale irrigation systems / edited by
Jonathan B. Mabry.
    p.   cm. — (Arizona studies in human ecology)
Includes bibliographical references (p.   ) and index.
ISBN 0-8165-1592-1 (acid-free paper)
1. Irrigation water — Developing countries — Management — Congresses.
2. Water-supply, Agricultural — Developing countries — Management —
Congresses.   3. Water rights — Developing countries — Congresses.
I. Mabry, Jonathan B., 1961–   .   II. Series.
HD1741.D44C36   1996
333.91'3'091724 — dc20   96-9958
CIP

British Library Cataloging-in-Publication Data
A catalogue record for this book is available from the British Library.

This volume is dedicated to

Robert McC. Netting

(1934–1995), a modern scholar of
the Enlightenment.

Contents

Acknowledgments

This book is the result of the efforts and encouragements of many people over several years. It grew out of the one-day workshop "Traditional Water Management in Comparative Perspective," which was held at the University of Arizona on April 29, 1990. Sponsors of the workshop included the Arizona State Museum; the Departments of Anthropology, Hydrology, Agricultural Resource Economics, Arid Lands Studies, and Near Eastern Studies; the College of Agriculture; and the Graduate College. This volume includes expanded versions of most of the papers presented, as well as several solicited manuscripts. David Groenfeldt, Robert Hunt, and Susan Lees served as discussants at the workshop. Their insights and suggestions aided the authors in the first drafts of their chapters. Robert Netting, Carol Gifford, Christine Szuter, Judith Wesley Allen, Linda Gregonis, and three anonymous reviewers suggested a number of useful revisions of later drafts. Christine Szuter and Judith Wesley Allen at the University of Arizona Press efficiently guided the volume through each stage of review and production. Ron Beckwith redrafted many of the figures, based on suggestions by Anne Keyl. William Doelle of the Center for Desert Archaeology also provided critical support for the completion of this volume.

Canals and Communities

The Ethnology of Local Irrigation

Jonathan B. Mabry

Irrigation, crucial for producing more than a third of today's total world harvest, has long been of practical and scholarly concern because of its proven ability to increase food supplies and its inherent logistical requirements for social coordination. Since the nineteenth century, archaeologists have unearthed evidence that irrigation supported the rise of many ancient states, and that, not occasionally, its mismanagement led to their collapse. Historians have chronicled how expanded irrigation systems provided infrastructures for the development of global colonial empires and the modern world economy. Engineers have distinguished diverse preindustrial water control technologies; geographers have traced the diffusion of those technologies across continents and oceans. Anthropologists have documented water management institutions ranging in formality from unwritten customs to codified laws, and varying in complexity from local irrigator alliances to centralized government bureaucracies. Economists have calculated irrigation's growing worth to the world's exponentially rising population, and international experts have recommended and helped implement numerous irrigation development projects to raise yields, incomes, and employment rates in developing countries. Politicians have, with increasing frequency, negotiated treaties to divide dwindling water resources among thirsty nations and peoples. Despite this enduring interdisciplinary interest, our knowledge about the social organization of irrigation is rather one-sided. Because of a common bias toward large, centrally controlled irrigation systems, the smaller, locally managed systems most familiar to anthropologists are less appreciated.

The Adaptability of Local Irrigation Systems

Archaeologists have long been impressed by the cases in antiquity when extensive and elaborate waterworks supported dense populations and advanced civilizations, and historians have focused on irrigation-based des-

Table 1.1 Numbers and sizes of small-scale, locally managed irrigation systems, 1980s, by country.

Country	Number of systems	Area (ha) (× 000)	% of total area irrigated
China	6,000,000+	—	—
India	400,000	8,400	20
Western United States	6,000+	6,200	34
Iran	28,000	5,300	95
Afghanistan	—	2,600	98
Bangladesh	—	2,300	95
Indonesia	5,000	2,200	40
Mexico	14,000	1,500	30
Nigeria	—	1,000	97
Philippines	5,500	850	53
Morocco	—	830	64
Japan	50,000	710	24
Sri Lanka	20,000	500	90
Nepal	—	400	44
South Korea	15,000	400	30
Yemen	—	280	92
Peru	3,000	250	16
Somalia	—	114	100
Lebanon	—	86	90

Sources: See text; additional sources: Baasiri and Ryan 1986; Barrow 1987; Groenfeldt 1989; Mandal 1987; Svendsen and Meinzen-Dick 1990; World Resources Institute 1994.

potic regimes ever since the 1930s when Karl A. Wittfogel (1957) asserted correlations between irrigation scale and political hierarchy. Development planners have typically assumed that centralized management is more efficient, and hydraulic engineers have tended to think that bigger is better. From the broader and longer perspectives of ethnology and archaeology, however, it is clear that large, centralized irrigation systems are relatively rare, short-lived, and limited to a narrow range of settings. Archaeological traces, historical documents, and ethnographic cases all indicate that the most common and enduring irrigation systems lie at the opposite end of the scale: They are small, locally built, farmer managed, and adapted to a wide variety of environments (table 1.1).

Some 28,000 *qanat* systems, the oldest built three thousand years ago, still tap groundwater to support oasis villages and irrigate more than five million hectares on the Iranian Plateau (Bonine, chapter 9). About 90 percent of the total irrigated area on the island of Sri Lanka is supplied by 20,000 village tanks, some more than two thousand years old (Medegama 1987; Stanbury, chapter 10). On the Indonesian island of Bali, careful flooding of terraced rice paddies by *subaks*—local irrigation communities—has been coordinated by networks of water temples for a millennium (Geertz 1980; Lansing 1987, chapter 7). Small canal systems up to four hundred years old irrigate about 400,000 hectares in the Himalayan foothills of Nepal (Upadhyay 1987). An estimated quarter of a million hectares of slopes terraced in pre-Hispanic times are still irrigated in the Andean highlands of Peru (Masson 1987; Gelles, chapter 5).

In the highlands of southern Arabia, carefully maintained terraces and diversions of pre-Islamic age still support over half of the population of the Arabian peninsula (Daverdisse 1980). Fifty thousand local irrigation systems, some in continuous operation since the early seventeenth century, represent almost a quarter of Japan's total irrigated area (Takeuchi 1980). In the Philippines, more than 5,500 *zanjeras,* indigenous local irrigation organizations first described by early seventeenth-century Spanish missionaries, currently irrigate about 850,000 hectares—almost two-thirds of the country's total irrigated area (World Bank 1982). European irrigation communities established more than five centuries ago still thrive on the Mediterranean coast of Spain (Glick 1970; Maass and Anderson 1978), and in the Swiss Alps (Netting 1974b, 1981). In the southwestern United States, the eastern Pueblo Indians manage thousand-year-old irrigation systems (Dozier 1970) alongside Hispanic irrigation communities established in the sixteenth century. About a thousand traditional Mexican-American water-user associations dating to the colonial period are extant in New Mexico alone (Crawford 1988).

In modern times, small-scale, community-based irrigation systems have demonstrated remarkable resilience in the face of significant shifts in the political economies they operate within, and in the context of rapid technological changes (Coward 1980:25). Institutions for local water management have survived not only outside the relatively restricted zones of large-scale irrigation projects in the developing world, but have also emerged independently in industrial countrysides. Today, it is estimated that 85 percent of the world's irrigated cropland still relies on small-scale, hand-built, gravity-flow canal systems, managed by local groups of farmers, organized into either independent irrigation communities or water-user associations at the base of larger systems (World Bank 1992). These surface systems, and also those locally managed systems based on tapping groundwater, diverting runoff, or retaining floodwaters, are known to range in size from less than 1 hectare to more than 4,000 hectares, but probably average less than 40 hectares (100 acres). In many countries they number in the thousands, and supply most or all of the irrigated cropland (table 1.1).

Small-scale systems operated by local irrigators are still used to grow most of the crops in Asia. In China, more than six million small tanks and reservoirs are village managed (Yu and Buckwell 1991). More than 15,000 systems fed by small reservoirs that average less than 50 hectares each are operated by local groups of farmers in Korea (Oh 1978). In the four southern states of India—Andhra Pradesh, Karnataka, Kerala, and Tamil Nadu—more than 125,000 community tanks supply 30 percent of the irrigated area (Sakthivadivel and Shanmugham 1987). Such systems play a similarly important role in the Middle East, Africa, and Latin America. In Morocco about two-thirds of the total irrigated area are watered by locally managed systems supplied by perennial flows and seasonal floods. The systems range in size from 50 to 3000 hectares and irrigate 830,000 hectares (Abdellaoui 1987; Welch, chapter 4). In Nigeria, about one million hectares are cultivated under some kind of traditional irrigation or flood-adapted cropping (Adams 1986). In valley wetlands in Zimbabwe, some 20,000 hectares of small garden plots called *bane,* which typically are less than a hectare each, are irrigated by hand-dug wells (Bell and Hotchkiss 1991). In the early 1980s there were almost 14,000 farmer-managed *unidades de riego* in Mexico, which ranged in size from 20 to 1,400 hectares and irrigated more than 1.5 million hectares (Hunt 1990; Sheridan, chapter 2). The significance of local systems is not restricted to the developing world, either. In 1978 there were 6,201 farmer-managed irrigation organizations in the seventeen western U.S. states, irrigating a total of 15.3 million acres (6.2 million ha), or more than a third of the total irrigated area (U.S. Department of Commerce 1979).

The widespread and sustained success of local irrigation systems raises broad questions about their origins, designs, functions, and responses: Under what kinds of economic and social conditions do they tend to develop? Are they sustained by a similar set of operational rules and rule-making mechanisms? Where is the locus of authority in decision making? What social institutions for resource management do they have in common? What are the typical effects of government intervention? Are there identifiable incentives for local participation in the development and management of new irrigation systems? Answers cannot be found through in-depth analysis of a single ethnographic case, but rather through cross-cultural comparisons—the ethnological approach to irrigation first employed by the French human geographer Jean Brunhes (1902, 1908).

Formation of Local Irrigation Organizations

The well-known and influential theory that all freely accessible resources will inevitably be overexploited, the "tragedy of the commons," holds that rational self-interest will always motivate individuals to take as much as possible, before others do the same, and that the only potential solutions are privatization or government regulation (Hardin 1968). But a wider, cross-cultural perspective reveals that competition for a limited and vital resource, such as water in dry lands or dry seasons, often results in "a social agreement that some resources are common property" (McCay and Acheson 1987:18). In these cases, the benefits and costs of use of common-pool resources are spread relatively equitably within a clearly bounded community of users. Based on comparative studies, a number of political scientists and anthropologists have concluded that local irrigator associations form for the same economic motivations as informal alliances among herders, timber cutters, and fishermen, and operate according to the same principles of common-property resource management (Coward 1979; Hunt 1986; Mahdi 1986; Cruz 1989; Ostrom 1990).

Local irrigator associations are not created by governments. They are grassroots organizations developed by local groups of farmers confronted with limited and variable water supplies, the problem of how to allocate them equitably and efficiently, and the decision of whether to invest labor and capital in infrastructures for water control. Through collective-choice processes the groups agree on rules that assign rights and responsibilities among the members, means of enforcing the rules, and mechanisms for resolving internal disputes. These are the basic structural arrangements that enable participants to commit to long-term, cooperative relations of production with each other (Tang 1992; Ostrom 1993).

Certain preconditions may foster the formation of local irrigation organizations, including relative evenness in economic status, significant environmental risk, and conflict-causing social competition. In a comparative study of ten different reservoir systems in India, it was found that local irrigation organizations are more likely to develop where there is low-to-moderate variance in the sizes of landholdings and incomes (Easter and Palanisami 1986). Environmental risks that can be at least partly mitigated by corporate institutional arrangements also encourage formation of local irrigation organizations. On the basis of his study of village irrigation systems in southern India, Robert Wade argues that collective water management is a social response to risky and unpredictable environments, conflicts over scarce resources, and potential crop losses: "Where risks associated with irrigation and common grazing are high, cultivators will straightforwardly come together to follow corporate arrangements designed to reduce those risks (Wade 1988:188).

Types of Local Irrigation Organizations

Local irrigation organizations can manage more than one kind of water control technology, and may form next to each other around canals, waterwheels, tanks, and tubewells in the same country or region. For example, in the Faiyum depression in Egypt, each secondary canal has a formally organized user group (*mesqa*) with an appointed water master (*ra'is*) who administers the rotational delivery of water shares in proportion to the sizes of land holdings, organizes maintenance, and settles disputes (Mehanna, Huntington, and Antonius 1984). In the Nile Delta and in other regions of Egypt where canal water flows below the level of the fields, irrigators are organized around the animal-powered waterwheels (*saqias*) that lift water to distribution canals. Ownership of each water wheel is divided into twenty-four equal shares (*qirat*) that correspond to turns in a time rotation. The largest landholder usually administers the operation and maintenance of the saqia and also arbitrates conflicts.

In addition, water management can be provided by more than one kind of local irrigation organization, from informal alliances of irrigators, to chartered water-user associations, to incorporated private enterprises. "Irrigation communities" and "water-user associations" can be distinguished—the former being autonomous, self-contained organizations that are supported by consensual, mutually beneficial institutional arrangements, while the latter are units within larger bureaucratic environments that require communication and coordination across organizational levels (Hunt 1989). A 1978 census identified several types of local irrigation organizations in the United States controlled and financed by user-operators, including "unincorporated mutuals," more formal "incorporated mutuals," and user-managed Bureau of Reclamation water projects (U.S. Department of Commerce 1979). Together, these represent 84 percent of all irrigation organizations in the United States. The remainder is comprised of systems constructed and operated by the Bureau of Reclamation, the Bureau of Indian Affairs, state and local governments, public corporations called "irrigation districts," and commercial enterprises.

Irrigation Scale and the Locus of Decision Making

Since Wittfogel introduced his model, anthropologists have debated about the relationship between the scale of irrigation systems and the complexity of managerial structures. Although most agree that hierarchical management institutions tend to correlate with large systems, they argue whether the relative centralization of decision making is more directly related to the size of the population supported by an irrigation system or to the size of its command area. Comparative data point to general correlations between management structure and these aspects of system size, but other studies raise the possibility that a more critical aspect of size is the number of irrigators that must coordinate their actions.

In a comparison of seventeen historic and contemporary case studies of irrigation systems, Kappel (1974) identified 5000 as the approximate upper limit of the number of people that can be supported by systems managed by local assemblies or councils; in this sample, central officials are necessary to effectively manage systems that support larger numbers of people. Among fifty cases (representing more than 100 irrigation systems), Uphoff (1986) found that those with command areas of 100 acres (40 hectares) or less tend to be managed by the entire group of irrigators, usually meeting in regular assemblies, while systems between 100 and 1000 acres (40 and 400 hectares) are usually administered by a central official, either elected by the irrigators or appointed by the state. Systems between about 1000 and 10,000 acres (400 and 4000 hectares) often have three levels of management institutions. Hunt (1988) compared fifteen irrigation systems in industrialized and developing countries, ranging between 700 and 30,000 hectares in size, and found no absolute relationship between system size and the structure of authority. But he predicted that any system larger than about 100 hectares has a "very high probability" of being managed by a "unified administrative authority structure" (Hunt 1988:347). This scale, 100 hectares (or 250 acres), is about the same upper limit for community-managed systems in forty-seven cases compared by Tang (1992), and may represent a cross-cultural threshold between democratic and bureaucratic management capacities.

These apparent correlations between scale and centralization may not be as straightforward as they seem, however. It may be that, because of the necessity to collect and process information from all parts of the system for efficient administration, it is the *number of irrigators,* rather than the total number of people supported or the scale of the total command area, that creates "scalar stress" in irrigation management. Psychological and sociological studies, supported by ethnological and archaeological data, suggest that, because of natural constraints on human information processing capabilities, difficulty in achieving consensus increases exponentially as organization size increases steadily (see Johnson 1982, and references therein). These studies show that consensus-based, task-oriented organizations have an average of six members (individuals or households), and an upper limit of about fifteen, above which management structures tend to either fission into multiple, smaller, independent units, or develop into hierarchical decision-making structures.

Alternatively, these may be "nonconsensual hierarchies," as Wittfogel predicts, or "consensual-sequential hierarchies," in which basal organizational units are integrated into larger units that can be managed by fewer decision makers, but in which lower order units can be only minimally coerced because they retain the option of fissioning from the system altogether—and occasionally do (Johnson 1982). The hierarchical system of common property in African flood recession agriculture, as described by Park (1992), may be an example of a third type that falls between the former and the latter: a "nonconsensual-sequential hierarchy" of kinship-based property rights that orders resource access through non-egalitarian, segmentary, and temporary political structures. In adaptation to high variability in seasonal flooding, these structures easily disintegrate during low flood years, with elite tribal segments retaining primary rights to smaller flooded areas, and segments with lower status shifting to mobile pastoralism.

The important conclusion based on ethnology is that, in addition to Wittfogel's model, there are a number of local organizational solutions to scalar stress in irrigation management, which may or may not include social inequality and political complexity. These alternate solutions are adapted to different environments and historical contexts and are derived from varied cultural traditions of resource ownership and use. Scarcity, it also turns out, may be just as important as scale in determining the locus of authority in irrigation management. On the basis of cross-cultural comparisons, Netting (1974a) suggests that, when it threatens widespread conflict that will seriously reduce the efficiency and utility of an irrigation system, water scarcity is also a stimulus for centralization of decision making. Lees (1974) reconstructs such a shift toward centralized control of water resources in the Valley of Oaxaca since the 1911 Mexican Revolution as a response to a receding water table.

Levels of Irrigation Management

Although small-scale irrigation systems with independent water sources and only a few irrigators can be managed by the irrigators themselves, many irrigators sharing common sources to irrigate larger areas must develop hierarchical decision-making institutions to manage more complex, integrated systems. In such structures the nature of tasks varies between levels, higher levels tend to operate more formally, and farmer participation in decision making occurs only at lower levels (Uphoff 1986). Multilevel irrigation systems can be divided into two types: state-managed systems administered directly by a government bureaucracy, and jointly managed systems administered by local associations in concert with government agencies (Coward 1980). In the latter, each level of organization, such as farmer associations, command area managers, and central project administrators, has unique roles, responsibilities, constituencies, and goals (Lusk 1991).

Within most jointly managed systems, water distribution is controlled by a bureaucracy of the state, but only until it reaches the level of local water-user associations. In southern Iraq during the early twentieth century, the distribution of water to the main canals, and their maintenance, was the responsibility of British colonial administrators and national bureaucrats, while the secondary canals were managed by local lineage groups organized into water-user associations (Fernea 1970). Local water-user associations also articulate with a national irrigation authority in Egypt, where the Ministry of Irrigation delivers water to each main canal in a rotation of turns, at which point it becomes the common property of all the farmers with plots accessible to the canal (Hunt 1986). In both these cases, local irrigation organizations exist at the base of a hierarchy of water management institutions.

Institutional Designs of Local Irrigation Organizations

Political scientists are particularly interested in the collective-action institutions of a self-governing irrigation system — the set of rules commonly known and used by participants to order repetitive, interdependent economic relationships with each other. These they consider to be a form of social capital that increases productive potential (Ostrom 1992). Some political scientists seek to categorize variability in the kinds of institutions that develop in these systems. Tang (1992), for example, suggests that these can be analytically divided into operational rules, such as rules concerning boundaries, allocation, inputs, and penalties, and collective-choice arrangements, such as mechanisms for monitoring compliance, adjudicating conflicts, enforcing decisions, and formulating and modifying operational rules. These two types of institutions are functionally related. Operational rules are constantly altered through collective-choice mechanisms to adjust to environmental variability, evolving farmer knowledge, and economic change.

Other political scientists are searching for the underlying logic of these local water management institutions. Through comparisons of community irrigation systems in Spain, the Philippines, Nepal, and other regions, Ostrom (1992) has identified a common set of design principles for management institutions in sustainable, self-governing irrigation systems: in order to maximize the efficiency and equitability of resource allocation, the boundaries of the service area must be clearly defined, and the group of individuals or households with water rights must be a closed set. Rules specifying the amount of water available to an irrigator must be congruent to the rules defining the required input levels of labor, materials, and capital in system operation and maintenance. Although membership in the user group is exclusive, the governing group that can modify the operational rules must be relatively inclusive, involving at least a majority of users. The operation and physical condition of an irrigation system should be mutually monitored by the irrigators themselves, or by an elected or appointed official who is directly accountable to them. Users who violate operational rules should be sanctioned by the other users, or by an official or council accountable to the users, in proportion to the seriousness and context of the offense. Users and their officials should have access to local mechanisms for rapid, affordable resolution of conflicts between users. The right of irrigators to devise their own water management institutions cannot be challenged by external government authorities. And finally, in larger systems, water management activities are organized into multiple, hierarchical levels nested within each other. Perhaps the most important of these design principles of successful irrigator associations (as for other local economic cooperatives) is boundary definition and maintenance, because without limitation of resource access to a bounded group there is open access, resulting in Hardin's "tragedy."

The solidarity and success of local irrigation communities and water-user associations is correlated with their ability to limit growth such that membership provides a secure share of water resources and protects them from "free-riding" outsiders. Membership can be based on proximity to other landholdings, on coresidence, or on structural position within a social hierarchy or physical delivery system. Principles of use rights based on multiple and overlapping social, ethnic, and religious criteria, as is the case for Muslim villagers, Mormon settlers, and Zionist kibbutzniks, clearly mark the limits of the irrigation community with intersecting criteria. Other typical criteria for defining membership are (1) ownership or leasing of land within the service area; (2) ownership or leasing of shares in water resources that are sometimes transferable independently of land holdings; (3) membership in a residential community or social organization; and (4) payment of taxes or fees for water use and delivery system maintenance (Tang 1992).

The term "irrigation community" describes the alliance of resource users—not a coresidence group. Because physical irrigation systems often crosscut several village territories, irrigation communities may be comprised of field neighbors from a number of different settlements (Coward 1979). A Balinese irrigator is a member of a particular village, for example, but he is also a member of a structurally separate subak, or even several, based on the location of his fields (Geertz 1980). In Ilocos Norte, Philippines, upstream zanjeras include residents of several villages, while downstream farmers may belong to more than one irrigation community to ensure that at least some plots receive water during dry years (Siy 1982). In other cases, due to logistics, membership in local cooperatives may be based on coresidence rather than kinship or proximity of properties. For instance, in central Luzon, the Philippines, residential neighbors often own scattered holdings in different areas. These neighbors form cooperative work groups (*kabesilyas*) that work the holdings at different times, thus balancing peaks in labor demand with the available work force (Francia 1988). In this way, peaks in labor demand are not out of balance with the available workforce.

In other local irrigation systems, rights to common-property water resources are associated with larger, external units of social hierarchies, such as segmentary tribal structures. The Erguita Berber tribe in the western highlands of Morocco considers water a collective good, and its management both links and opposes the various social segments that compose the tribe (Mahdi 1986). Water is owned collectively by different tribal fractions occupying groups of villages, though some villages are composed of more than one fraction living in separate sections of the same settlement. Within each village (*douar*), water rights are divided among all of the lineages (*ikhs*) represented, which in turn distribute use rights among their individual households (*takatines*). Measured either in units of time or volume, water rights may not be sold or otherwise alienated from the lineage to which they ultimately belong, and into which households coalesce during conflicts with other lineages.

Management Tasks of Local Irrigation Organizations

Although political scientists and cultural ecologists have identified some of the underlying motivations for and principles of collective action and cooperative resource management, social scientists studying the relationships between irrigation and rural social organization have recognized between five and twelve basic activities, or tasks, necessary for successful local irrigation management (table 1.2). Uphoff (1986) categorizes these in terms of activities related to water use (acquisition, allocation, distribution, drainage), water control structures (design, construction, operation, maintenance), and organization (decision making, resource mobilization,

Table 1.2   Basic tasks of local water management, based on ethnological comparisons.

Hunt and Hunt 1976	Coward 1979	Kelly 1983	Freeman and Lowdermilk 1985	Uphoff 1986
Acquisition	Acquisition	Aquisition	—	Acquisition
Allocation	Allocation	Allocation	Allocation	Allocation
—	—	Delivery	—	Distribution
—	—	Drainage	Drainage	Drainage
—	—	—	—	Design
Construction	—	Construction	Construction	Construction
—	—	Operation	—	Operation
Maintenance	Maintenance	Maintenance	Maintenance	Maintenance
Decision-making	—	—	—	Decision-making
Resource mobilization	Resource mobilization	—	—	Resource mobilization
—	—	—	—	Communication
Conflict resolution	Conflict resolution	Conflict resolution	Conflict resolution	Conflict management

Source: Uphoff 1986

communication, conflict management), and he relates these categories to the fundamental factors of production considered by economists: land (or resources), capital (infrastructure investments), and labor (division of management effort). The following discussion refers to ethnographies of irrigation communities to discuss some cross-cultural patterns in *resource allocation principles* and *conflict resolution mechanisms* — perhaps the most important tasks faced by locally organized irrigators everywhere.

Allocation

In many cultures, long-fallow agricultural lands, pasturage, woodlands, fishing areas, and other diffuse, extensive, and marginally productive resources are communally held, but factors such as increased competition for resources under population pressure or investments of capital and labor that increase productivity (irrigation, terracing, or planting of orchards, for example), usually lead to individual property rights and inheritance rules (Netting 1982). For these same reasons, in almost every locally managed irrigation system the water resource itself, the physical means to capture and distribute it, and the labor necessary to construct and maintain the water control infrastructure are held in common by an association of users, but the water is used to irrigate private, transferable, and heritable plots, the produce of which belongs to individuals or households. The appropriations from common property are invariably private (Netting 1993a, 1993b).

This pattern is illustrated in the eastern Jordan Valley, where the internationally funded, large-scale East Ghor Canal Project has bypassed a few old villages because of the uneven topography. Here, between areas irrigated by the modern canal, Arab communities continue to operate small-scale canal systems according to a mixture of private and collective property rights that date to medieval times (Mabry 1991). Water resources are ultimately owned by the communities, and allocation is administered by councils of elders called "the faces of the tribes" (*el-wozhuh el-qabila*), representing each patrilineal clan (*hamula*) in the community. The entire flows of perennial springs and streams are distributed in rotation for set units of time called "seasons" (*fusul*), according to inherited shares owned by each clan. Water rights may be sold with or without land, but they may not be sold outside of the clan. Because of the higher investments and inputs into irrigated land, mostly planted in vegetables for local markets, rights in this land have shifted in large part to the individual household, reducing the degree of common property control by each clan. Before British colonial administrators privatized communal landholdings (*masha'a* lands) during the 1930s, it was the lower value rainfed lands surrounding the irrigated zone that were traditionally held in common, with use rights rotated among the clans from year to year to ensure an even distribution of good, bad, and average yields of cereal crops for domestic consumption (Atran 1986). This example also demonstrates that the dichotomy of common property and private property is seldom simple. Traditional tenure systems are usually complex mixtures of rights that serve to redistribute resources within a community.

As in some other systems of common property resource management, local irrigator associations may allocate irrigated fields in such a way as to reduce risks and level disparities within the group. In *zanjera* irrigation communities in the Ilocos Norte region of the Philippines, land is divided into several large sections, and every farmer has a plot in each. With this system, each member has access to the best plots near the head of the irrigation system, while during dry years, shortfalls are shared equitably by not irrigating lands near the tail of the system (Siy 1982). The villagers of Pul Eliya in Sri Lanka also divide their land into sections, with each irrigator maintaining a portion of his holdings in each zone, thus avoiding the risks and conflicts of a system where some irrigators' entire holdings are at the head of the system where there is the most available water, and others' are all at the tail end where water availability is tenuous (Leach 1961). Due to varying concepts of resource ownership and proper use, however, equitable distribution of risks and benefits is not a universal goal of water allocation in indigenous systems.

Water scarcity often clarifies the principles of customary and formalized systems of water allocation. Under Yemenite systems, one of two common principles of allocation in the Islamic world, water is a commodity potentially separate from land, and shortages are absorbed by users with secondary rights; in Syrian systems, water rights are assigned on the basis of types of land ownership, and average and reduced flows are divided proportionally among all users (Glick 1970). Among Berber tribes in eastern Morocco using a sequential, Syrian type of allocation, water sources are considered collective property and use rights are attached to specific parcels of land (Welch, chapter 4). During shortages, the reduced volume of streamflow is distributed in proportion to the relative sizes of land holdings. By contrast, under a Yemenite type of system in this same region, water rights are measured in time units, are owned separately from land, and there is increased competitive speculation in water rights during shortages.

More formal sets of laws defining principles of water ownership and use have been developed within the five major legal traditions of world history—Roman, Islamic, English, Chinese, and Hindu (Radosevich 1987), and specific historical and environmental circumstances have often led to complex cultural overlays of legal principles in the same regions. In southern Arabia, pre-Islamic customs are blended with Islamic laws in the management of flood irrigation systems (Maktari 1971) and filtration galleries (*aflaj*) that tap groundwater sources (Wilkinson 1977). In Spain, water laws developed as a mixture of late medieval Iberian laws and earlier Roman and Islamic legal traditions (Glick 1970). The use of the "riparian-rights principle" of water ownership in the humid eastern United States derived from English common law during the colonial period, but the principle of "prior appropriation," which grants primary rights to those who first use the water, developed in the arid West during the competitive mining boom of the mid- to late nineteenth century (Norvelle 1988). Bennett (1974) notes that the riparian-rights principle resembles the Yemenite system, while the doctrine of prior appropriation is equivalent to the Syrian system. These two principles of water allocation, although not the only ones, appear to represent common alternatives.

Conflict Resolution

Social cooperation is required to manage the hydraulic commons, but capital and labor investments in waterworks, irrigation, and intensive crop production create a sharply bounded zone of higher land value, resulting in increased competition and potential conflict over water rights. In *comunes de agua* in Sonora, Mexico, the contrast between the equity of communal regulation of surface water resources and the inequity of unlimited individual use of groundwater (a dichotomy derived from distinctions in traditional Hispanic water law) is a significant source of conflict in the community, and an example of the common tension between cooperation and competition in local water management (Sheridan, chapter 2). Differential access to higher value fields in *dhasheeg* flood basins in the Jubba Valley of Somalia is both a source of conflict and a means of maintaining the traditional stratification of power (Besteman, chapter 3).

To resolve such conflicts, an elected judge, a council of representatives, or some other institutionalized mechanism is necessary. Water judges (*cequieres*) are appointed in the *huertas* of Valencia, Spain, to monitor the distribution of water, patrol for water theft, arbitrate disputes, and levy fines for infractions of the rules (Glick 1970). Serious conflicts are referred to a special water court (*tribunal de las aguas*). In other parts of the world, water conflicts between irrigator groups are handled by state judicial systems. Archives show that conflicting water claims between Indian villages, Catholic missions, and military garrisons in "New Spain" were ultimately settled in the crown-administered colonial courts (Meyer 1984).

It is in the interest of irrigators to settle conflicts among themselves, as arbitration by outsiders represents a loss in local autonomy. But when local interests become too factionalized, only outside mediators can solve the structural impasse. A water dispute that began in 1918 between two political alignments within the same clan of the Nahid tribe of southern Arabia could not be internally resolved because neither faction's chieftain could serve as an objective arbitrator, their customary role in disputes between clans and with members of other tribes. This paralysis was not resolved until 1936, when an agreement was forged by a British colonial administrator (Hartley 1961, as cited in Millon 1962).

Water theft is a constant source of tension among wet rice farmers in Taiwan, forcing them to form larger, intercommunity irrigation associations with the authority to monitor water turns (VanderMeer 1971). Dispute resolution among Berber irrigation communities in central Morocco may be best understood in terms of a "hierarchy of responses" to stress that encourages resolution at lower levels (Welch, chapter 4). Local water judges (*jārí*) serve each association of water users along a secondary canal, but decisions may be appealed first to the lineage council, then to the tribal council, then finally to government authorities. Each step upward in the hierarchy of options, however, represents a decrease in local autonomy and an increase in social cost.

Effects of State Intervention in Local Irrigation

Because of their capacity to survive outside of state structures and market economies, rural irrigation communities can endure almost any political collapse and economic crash caused by unpopular policies or inept management by urban-based regimes. For their own political ends, central governments often seek to undermine the autonomy of irrigation communities by appropriating control over water resources, forcibly integrating them into larger irrigation systems, and increasing their dependence on the state through construction of new waterworks that are beyond the management capabilities and capital resources of rural groups (Lees 1974; Enge and Whiteford 1989). The conflict between the modernizing state and traditional irrigation communities can be seen in the highlands of Peru in terms of the state's "rationalized model" and the community's "ritualized model" of water management, and between the simultaneous tendencies for bureaucratic intervention and peasant resistance (Gelles, chapter 5).

The result of such state intervention is a recipe for instability: integration into larger, centralized systems decreases capacities to respond quickly and appropriately to local problems and increases the effect of distant disturbances on local subsystems. History is familiar with this trajectory. In Mesopotamia and other regions during ancient times, large-scale irrigation systems tended to collapse because they relied on centralized decision making insulated from local pressures, they undermined their long-term resource base through production maximizing, and they displaced local resource management institutions that could better adapt to changing conditions (Adams 1978). Recent history has also shown that, although locally organized water-user associations and other grass-roots agricultural cooperatives often succeed, those that are imposed from above rarely do (Bennett 1983; Hyden 1988). Irrigator associations organized from the top down in Tunisia, for example, have collapsed or performed poorly because they are socially and economically isolated from traditional rural organizations, and because their members have no permanent rights to land and water (Hopkins 1988). Similar effects have been documented for federally sponsored irrigator associations on Native American lands in the southwestern United States (Bentley 1987).

Hunt (1990) argues that the main reason government-created water-user associations fail is a lack of sovereignty over resources and an absence of local recognition of use rights: "In these cases the state is trying to create these corporate groups. The danger with such attempts is that only external jurality is being created. (. . . where the group has jural standing with entities outside the group). In many cases this external jurality vanishes as soon as the state's representatives vanish. . . . My argument is that farmers are unlikely to participate in maintenance activities without clear benefits, which I believe are principally found in control over the acquisition and allocation of the water, both of which are a function of the internal jurality of the corporate group managing the commons (Hunt 1990:148)."

Incentives for Local Participation in Development

Comparing government administered and farmer managed irrigation systems in Mexico, Hunt (1990) concluded that the degree of control over resources, not size, is the key difference in determining the level of farmer participation in large-scale irrigation management and development. Other research suggests that the level of participation is also related to farmers' goals of minimizing risk, securing a role in decision making that affects them, and influencing the direction of development to best fit their needs (Parlin and Lusk 1991). In a study of nearly 500 farmers in the 125,000-hectare Gal Oya irrigation scheme in Sri Lanka, researchers found that farmers are more willing to participate in management where water is relatively scarce, rather than where it is absolutely scarce or abundant (Uphoff, Wickramasinghe, and Wijayaratna 1990). Farmers in the middle of the system were more likely to form user committees than to delegate authority to a traditional irrigation headman (*vidane*) or to a government officer because there was more variance in the water supply and more perceived conflicts in the middle than in the head or tail ends of the system. Thus they had more to gain from participation in management activities that stabilized their shares of the supply, minimized conflict, and gave them a voice in decision making that can directly benefit them. In a study of irrigation systems in the hills of Nepal, it was also found that when farmers banded together to acquire water supplies, they were more likely to participate in allocation, operation, maintenance, decision making, and conflict resolution (Martin and Yoder 1987).

Local irrigation communities in the Ziz Valley of southern Morocco also play a large role in development of small- and medium-scale waterworks. Participation insures that projects are tailored to local needs, and guarantees the rights of farmers to control their operation (Mabry, Ilahiane, and Welch 1991). In the planning stages, administrators and engineers from the provincial government and the regional office of the national Agricultural Ministry meet with local groups to inform them about the available budget and to find out their priorities. During construction, the farmers volunteer most or all of the necessary labor, while the local government and the Agricultural Ministry design the structures and provide the materials. Large savings typically result from the contribution of labor, and the remaining funds are invested in another local project, from which similar savings provide the seed for still another project, and so on in a continuing chain. Management of the new works is the responsibility of the local groups that construct them, which has had better results than "turn-key" projects planned and built entirely by the government and then turned over to farmers to run. In the Philippines, decentralization of irrigation management to local *zanjeras* has resulted in increased efficiency in water distribution and a higher rate of compliance to rules (Korten and Siy 1988). A new approach to irrigation development in Sri Lanka, in which local groups are involved with planning and management in ways modeled after indigenous tank irrigation systems, is also proving successful (Stanbury, chapter 10).

Ultimately, *who* specifically benefits from government-administered irrigation development schemes is an important factor in their long-term success and acceptance. Modern development of water resources has often served to widen the gap between indigenous elites and the poorest farmers. Large landholders are favored by tubewell development in the Cape Verde Islands "because of the advantages that automatically accrue to wealth, size, and status" (Langworthy and Finan, chapter 8). The traditional Tswana elite class in Botswana, who controlled cattle production during the colonial period, later formed exclusionary syndicates to invest in tubewells for watering their herds (Peters 1988). With the support of government land privatization programs, they have claimed ownership of the grazing lands surrounding the wells—lands previously held in common by the entire tribe. Among the Sonjo in Tanzania, prestigious and powerful positions in the water management hierarchy may be bought or sold, but they are normally inherited within elite lineages (Gray 1963). The high costs of irrigation in government-developed schemes in Tunisia have further marginalized the poorer farmers to the benefit of richer ones (Salem-Murdock 1990).

Ethnology and Local Irrigation Organizations

The comparative method of ethnology helps researchers to identify some of the significant regularities in the formation and functioning of successful local irrigation organizations. These grassroots, small-scale social formations are self-governing, collective-choice institutions for managing commonly held resources—water, storage and delivery systems, and labor—for irrigated agricultural production. They tend to form under conditions of relative resource scarcity, economic risk, and social competition. These organizations are composed of farmers allied by the necessity of equitably dividing a limited water supply, and the need to pool labor and capital for construction and maintenance of irrigation infrastructures. They are held together by shared ecological risks, mutual economic interests, and collective investments in the means of production. They must handle a similar set of tasks, and the locus of decision making is largely determined by the scale of management, in terms of the size of the system and the number of irrigators. They are sustained by compliance with rules that spread risk, level disparities, minimize conflict, and define boundaries. Local irrigation organizations are exclusive in membership, territorial in defense of resources, resistant to outside intervention, and resilient in the face of change. Restricted membership, however, sometimes includes local elites and excludes less powerful classes, and members' shares in common resources are often proportional to their relative power, wealth, and status. Outside intervention that reduces local autonomy is usually resisted, but acquisition of new water resources, stabilization of existing supplies, reduction of risk, inclusion in decision making, and equitable distribution of benefits are strong incentives for local participation in irrigation development projects.

Ethnological insight derives from the details of specific cases as well as from broad comparisons. This book—a collection of case studies and comparative essays covering a range of environments, cultural traditions, and historical contexts—is about how local groups manage water resources to augment crop production through social institutions adapted to specific settings and derived from unique cultural histories. Common threads connecting the case studies and essays include incentives for cooperation, operational rules and collective-choice arrangements, allocation principles and conflict-resolution mechanisms, links between scale and complexity, and relationships between autonomy, diversity, and stability. The book provides sources for those investigating the cultural ecology of irrigated agriculture, the ethnology of cooperative social formations and collective-choice institutions, and the sociology of rural development. It also provides examples and generalizations of the cross-cultural characteristics of sustainable water management and intensive agriculture.

Acknowledgments

Robert Netting, Christine Szuter, John Welch, and Greg McNamee commented on drafts of this chapter and helped the author simplify comparisons and clarify conclusions. Any remaining muddy thinking or writing is the fault of the author.

References Cited

Abdellaoui, R. M.
  1987   Small-scale Irrigation Systems in Morocco: Present Status and Some Research

Issues. In *Public Intervention in Farmer-Managed Irrigation Systems,* 165-73. Digana, Sri Lanka: International Irrigation Management Institute.

Adams, Robert McC.

1978   Strategies of Maximization, Stability, and Resilience in Mesopotamian Society, Settlement, and Agriculture. *Proceedings of the American Philosophical Society* 122(5):329-35.

Adams, W. M.

1986   Traditional Agriculture and Water Use in the Sokoto Valley, Nigeria. *The Geographical Journal* 152:30-43.

Atran, Scott

1986   Hamula Organisation and Masha'a Tenure in Palestine. *Man* 21(2):271-95.

Baasiri, Mouin, and John Ryan

1986   *Irrigation in Lebanon: Research, Practices and Potential.* Beirut: National Council for Scientific Research.

Barrow, C.

1987   *Water Resources and Agricultural Development in the Tropics.* Harlow: Longman Scientific and Technical.

Bell, Morag, and Patricia Hotchkiss

1991   Garden Cultivation, Conservation and Household Strategies in Zimbabwe. *Africa* 58:315-36.

Bennett, John

1974   Anthropological Contributions to the Cultural Ecology and Management of Water Resources. In *Man and Water: The Social Sciences in Management of Water Resources,* ed. L. Douglas James, 34-81. Lexington: University of Kentucky Press.

1983   Agricultural Cooperatives in the Development Process: Perspectives from Social Science. *Studies in Comparative International Development* 18:3-68.

Bentley, Jeffery W.

1987   Water Harvesting on the Papago Reservation: Experimental Agricultural Technology in the Guise of Development. *Human Organization* 46(2):141-46.

Brunhes, Jean

1902   *L'Irrigation: Ses Conditions Geographiques, ses Modes et son Organisation.* Paris.

1908   *Les Différents Systèmes d'Irrigation.* Brussels: Bibliotèque Coloniale Internationale.

Coward, E. Walter, Jr.

1979   Principles of Social Organization in an Indigenous Irrigation System. *Human Organization* 38:28-36.

1980   Irrigation Development: Institutional and Organizational Issues. In *Irrigation and Agricultural Development in Asia: Perspectives from the Social Sciences,* ed. E. W. Coward, Jr., 15-27. Ithaca: Cornell University Press.

Crawford, Stanley

1988   *Mayordomo: Chronicle of an Acequia in Northern New Mexico.* New York: Doubleday.

Cruz, C.

1989   Water as Common Property: The Case of Irrigation Water Rights in the Philippines. In *Common Property Resources: Ecology and Community-Based Sustainable Development*, ed. F. Berkes, 218–35. London: Belhaven.

Daverdisse, G.

1980   *Aspects Techniques et Sociaux de l'Irrigation en Republique Arabe du Yemen.*
Paris: Institut de la Recherche Agronomique.

Dozier, Edward

1970   *The Pueblo Indians of North America.* New York: Holt, Rinehart and Winston.

Easter, K. William, and K. Palanisami

1986   *Tank Irrigation in India and Thailand: An Example of Common Property Resource Management.* Staff paper, Department of Agricultural and Applied Economics, University of Minnesota, Minneapolis.

Enge, Kjell I., and Scott Whiteford

1989   *The Keepers of Water and Earth: Mexican Rural Social Organization and Irrigation.* Austin: University of Texas Press.

Fernea, Robert

1970   *Shaykh and Effendi: Changing Patterns of Authority Among the El Shabana of
Southern Iraq.* Cambridge, Mass.: Harvard University Press.

Francia, J.

1988   An Economic Analysis of Co-operative Labour among Philippine Rice Farmers.
In *Who Shares? Co-operatives and Rural Development*, ed. D. Attwood and
B. Baviskar, 259–81. Delhi: Oxford University Press.

Freeman, David M., and Max L. Lowdermilk

1985   Middle-level Organizational Linkages in Irrigation Projects. In *Putting People
First: Sociological Variables in Rural Development*, ed. Michael Cernea, 91–118.
Washington, D.C.: World Bank-Oxford University Press.

Geertz, Clifford

1972   The Wet and the Dry: Traditional Irrigation in Bali and Morocco. *Human
Ecology* 1(1):23–39.

1980   Organization of the Balinese Subak. In *Irrigation and Agricultural Development
in Asia: Perspectives from the Social Sciences*, ed. E. W Coward, Jr., 70–90. Ithaca:
Cornell University Press.

Glick, Thomas

1970   *Irrigation and Society in Medieval Valencia.* Cambridge: Harvard University
Press.

Gray, Robert

1963   *The Sonjo of Tanganyika: An Anthropological Study of an Irrigation-Based
Society.* London: Oxford University Press.

Groenfeldt, David

1989   Indigenous Irrigation Knowledge and Agricultural Development. Paper presented at the Annual Meeting of the American Anthropological Association.
Washington, D.C.

Hardin, Garrett

1968   The Tragedy of the Commons. *Science* 162:1243–48.

Hartley, John A.

1961   *The Political Organization of an Arab Tribe of the Hadhramaut.* Ph.D. dissertation, London School of Economics, London.

Hopkins, Nicholas S.

1988   Co-operatives and Non-cooperative Sector in Tunisia and Egypt. In *Who Shares? Co-operatives and Rural Development,* ed. D. Attwood and B. Baviskar, 211–30. Delhi: Oxford University Press.

Hunt, Robert C.

1986   Canal Irrigation in Egypt: Common Property Management. In *Proceedings of the Conference on Common Property Management, 1985,* 199–214. Washington, D.C.: National Academy Press.

1988   Size and Structure of Authority in Canal Irrigation. *Journal of Anthropological Research* 44(4):335–355.

1989   Appropriate Social Organization? Water Users Associations in Bureaucratic Canal Irrigation Systems. *Human Organization* 48:79–90.

1990   Organizational Control Over Water: The Positive Identification of a Social Constraint on Farmer Participation. In *Social, Economic, and Institutional Issues in Third World Irrigation Management,* ed. R. K. Sampath and R. A. Young, 141–54. Boulder, Colo.: Westview Press.

Hunt, Robert C., and Eva Hunt

1976   Canal Irrigation and Local Social Organization. *Current Anthropology* 17(3): 389–411.

Hyden, G.

1988   Approaches to Co-operative Development: Blueprint versus Greenhouse. In *Who Shares? Co-operatives and Rural Development,* ed. D. Attwood and B. Baviskar, 149–71. Delhi: Oxford University Press.

Johnson, Gregory A.

1982   Organizational Structure and Scalar Stress. In *Theory and Explanation in Archaeology: The Southampton Conference,* ed. C. Renfrew, M. J. Rowlands, and B. A. Segraves, 389–421. New York: Academic Press.

Kappel, Wayne

1974   Irrigation Development and Population Pressure. In *Irrigation's Impact on Society,* ed. T. E. Downing and McG. Gibson, 159–67. Anthropological Papers No. 25. Tucson: University of Arizona Press.

Kelly, William

1983   Concepts in the Anthropological Study of Irrigation. *American Anthropologist* 85:880–86.

Korten, Frances F., and Robert Y. Siy, Jr. (eds).

1988   *Transforming a Bureaucracy: The Experience of the Philippine National Irrigation Administration.* West Hartford, Conn.: Kumarion Press.

Lansing, J. Stephen
   1987   Balinese "Water Temples" and the Management of Irrigation. *American Anthro-
           pologist* 89(2):326–41.
Leach, Edmund R.
   1961   *Pul Eliya: A Village in Ceylon.* Cambridge: Cambridge University Press.
Lees, Susan H.
   1974   The State's Use of Irrigation in Changing Peasant Society. In *Irrigation's Im-
           pact on Society,* ed. T. E. Downing and McG. Gibson, 123–28. Anthropological
           Papers No. 25. Tucson: University of Arizona Press.
Lusk, Mark W.
   1991   Irrigation Experience Transfer: The Social Dimension. In *Farmer Participation
           and Irrigation Organization,* ed. B. W. Parlin and M. W. Lusk, 69–96. Boulder,
           Colo.: Westview Press.
Maass, Arthur, and Raymond L. Anderson (eds.)
   1986   . . . *and the Desert Shall Rejoice: Conflict, Growth, and Justice in Arid Environ-
           ments.* Cambridge, Mass.: MIT Press.
Mabry, Jonathan B.
   1991   The Seasons of Water: Traditional Irrigation and Development in Jordan and
           Morocco. Paper presented at the 24th Annual Chacmool Conference, Calgary.
Mabry, Jonathan B., Hsain Ilahiane, and John R. Welch
   1991   *Rapid Rural Appraisal of Moroccan Irrigation Systems: Methodological Lessons
           from the Pre-Sahara.* Report submitted to United States Agency for Interna-
           tional Development-Morocco, Rabat.
Mahdi, M.
   1986   Private Rights and Collective Management of Water in a High Atlas Berber
           Tribe. In *Proceedings of the Conference on Common Property Resource Manage-
           ment, 1985,* 181–97. Washington, D.C.: National Academy Press.
Maktari, A.
   1971   *Water Rights and Irrigation Practices in Lahj: A Study of the Application of Cus-
           tomary and Shari'ah Law in Southwest Arabia.* Oriental Publications No. 21.
           Cambridge: Cambridge University Press.
Mandal, M. A. S.
   1987   Development of Small-scale Lift Irrigation in Bangladesh. In *Public Intervention
           in Farmer-Managed Irrigation Systems,* 131–47. Digana, Sri Lanka: International
           Irrigation Management Institute.
Martin, Edward, and Robert Yoder
   1987   *Institutions for Irrigation Management in Farmer-Managed Systems: Examples
           from the Hills of Nepal.* Research Paper No. 5. Digana, Sri Lanka: International
           Irrigation Management Institute.
Masson, Luis
   1987   La ocupación de andenes en Perú. *Pensamiento Iberoamericano* 12(2):179–200.
Medegama, Jaliya
   1987   State Intervention in Sri Lanka's Village Irrigation Rehabilitation Program. In

Public Intervention in Farmer-Managed Irrigation Systems, 215–32. Digana, Sri Lanka: International Irrigation Management Institute.

Mehanna, S., R. Huntington, and R. Antonius

1984 *Irrigation and Society in Rural Egypt.* Cairo Papers in Social Science, Vol. 7, Monograph 4.

Meyer, Michael

1984 *Water in the Hispanic Southwest: A Social and Legal History, 1550–1850.* Tucson: University of Arizona Press.

Millon, Réné

1962 Variations in Social Responses to the Practice of Irrigated Agriculture. In *Civilization in Arid Lands,* ed. R. B. Woodbury, 56–88. Anthropological Papers No. 62. Salt Lake City: University of Utah Press.

Netting, Robert McC.

1974a Agrarian Ecology. *Annual Review of Anthropology* 3:21–56. Palo Alto: Annual Reviews.

1974b The System Nobody Knows: Village Irrigation in the Swiss Alps. In *Irrigation's Impact on Society,* ed. T. E. Downing and McG. Gibson, 67–75. Anthropological Papers No. 25. Tucson: University of Arizona Press.

1981 *Balancing on an Alp.* Cambridge: Cambridge University Press.

1982 Territory, Property and Tenure. In *Behavioral and Social Science Research: A National Resource,* ed. R. McC. Adams, N. Smelser, and D. Treiman, 446–502. Washington, D.C.: National Academy Press.

1993a Communal Labor as a Common Property Resource: How Smallholders Work Together. Paper presented at IASCP Fourth Annual Common Property Conference, Manila, Philippines.

1993b Unequal Commoners and Uncommon Equity: Property and Community Among Smallholder Farmers. Paper presented at "Heterogeneity and Collective Action," Workshop in Political Theory and Policy Analysis, Indiana University, Bloomington, Indiana.

Norvelle, Michael E.

1988 The Development of Water Law in the Arid Lands of the Middle East and the United States: A Comparative Study. In *Arid Lands Today and Tomorrow,* ed. E. E. Whitehead, C. F. Hutchinson, B. N. Timmermann, and R. G. Varady, 289–303. Boulder, Colo.: Westview Press.

Oh, Ho-Sung

1978 Customary Rules of Water Management for Small Irrigation Reservoirs in Korea. *Journal of Rural Development* 1:96–110.

Ostrom, Elinor

1990 *Governing the Commons: The Evolution of Institutions for Collective Action.* Cambridge: Cambridge University Press.

1992 *Crafting Institutions for Self-Governing Irrigation Systems.* San Francisco: ICS Press.

1993 Constituting Social Capital and Collective Action. Paper presented at "Heterogeneity and Collective Action," Workshop in Political Theory and Policy Analysis, Indiana University, Bloomington.

Park, Thomas K.

1992   Early Trends Toward Class Stratification: Chaos, Common Property, and Flood Recession Agriculture. *American Anthropologist* 94(1):90–117.

Parlin, Bradley W., and Mark W. Lusk

1991   *Farmer Participation and Irrigation Organization.* Boulder, Colo.: Westview Press.

Peters, P.

1988   The Ideology and Practice of Tswana Borehole Syndicates: Co-operative or Corporation? In *Who Shares? Co-operatives and Rural Development,* ed. D. Attwood and B. Baviskar, 23–45. Delhi: Oxford University Press.

Radosevich, George E.

1987   Water Policy and Law: The Missing Link in Food Production. In *Water and Water Policy in World Food Supplies,* ed. W. R. Jordan, 185–91. College Station: Texas A & M University Press.

Sakthivadivel, R., and C. R. Shanmugham

1987   Issues Related to Interventions in Farmer-managed Irrigation: Rehabilitation of a Tank Irrigation System. In *Public Intervention in Farmer-Managed Irrigation Systems,* 113–30. Digana, Sri Lanka: International Irrigation Management Institute.

Salem-Murdock, Muneera

1990   Household Production Organization and Differential Access to Resources in Central Tunisia. In *Anthropology and Development in North Africa and the Middle East,* ed. M. Salem-Murdock and M. M. Horowitz, with M. Sella, 95–125. Boulder, Colo.: Westview Press.

Siy, Robert Jr.

1982   *Community Resource Management: Lessons from the Zanjera.* Quezon City: University of the Philippines Press.

Svendsen, Mark, and Ruth Meinzen-Dick

1990   *Garden Irrigation: The Invisible Sector.* Washington, D.C.: International Food Policy Research Institute.

Takeuchi, Satoru

1980   Japan. In *Farm-level Water Management in Selected Asian Countries,* ed. H. Takeuchi, 76–90. Tokyo: Asian Productivity Organization.

Tang, Shui Yan

1992   *Institutions for Collective Action: Self-Governance in Irrigation.* San Francisco: ICS Press.

U.S. Department of Commerce

1979   *Census of Irrigation Organizations (1978).* Washington, D.C.: Bureau of the Census.

Upadhyay, S. B.

1987   Public Intervention in Farmer-managed Irrigation Systems in Nepal. In *Public*

Intervention in Farmer-Managed Irrigation Systems, 233–36. Digana, Sri Lanka: International Irrigation Management Institute.

Uphoff, Norman

1986   *Improving International Irrigation Management with Farmer Participation: Getting the Process Right.* Boulder, Colo.: Westview Press.

Uphoff, Norman, M. L. Wickramasinghe, and C. M. Wijayaratna

1990   "Optimum" Participation in Irrigation Management: Issues and Evidence from Sri Lanka. *Human Organization* 49(1):26–40.

VanderMeer, C.

1971   Water Thievery in a Rice Irrigation System in Taiwan. *Annals of the Association of American Geographers* 61:156–79.

Wade, Robert

1988   *Village Republics: Economic Conditions for Collective Action in South India.* Cambridge: Cambridge University Press.

Wilkinson, J.

1977   *Water and Tribal Settlement in Southeast Arabia: A Study of the Aflaj of Oman.* Oxford: Oxford University Press.

Wittfogel, Karl A.

1957   *Oriental Despotism: A Comparative Study of Total Power.* New Haven, Conn.: Yale University Press.

World Bank

1982   *Philippines Communal Irrigation Development Project.* Washington, D.C.: World Bank.

1992   *World Development Report 1992: Development and the Environment.* Washington, D.C.: World Bank and Oxford University Press.

World Resources Institute

1994   *World Resources, 1993–94.* New York: Oxford University Press.

Yu, Chen Liang, and Allan Buckwell

1991   *Chinese Grain Economy and Policy.* Wallingford: C. A. B. International.

Part 1

Patterns of Cooperation and Conflict in Local Irrigation

2

La Gente Es Muy Perra
Conflict and Cooperation over Irrigation Water in
Cucurpe, Sonora, Mexico

Thomas E. Sheridan

In *Mayordomo: Chronicle of an Acequia in Northern New Mexico,* Stanley
Crawford writes about his year as a ditch-boss in a largely Hispanic community:

> Next to blood relationships, which rule the valley, come water re-
> lationships. The arteries of ditches and bloodlines cut across each
> other in patterns of astounding complexity. Some families own prop-
> erties on two or three of the valley's nine ditches. You can argue that
> the character of a man or woman can be as much formed by genetic
> and cultural material as by the location of their garden or chile patch
> along the length of a ditch, toward the beginning where water is plen-
> tiful or at the tail end where it will always be fitful and scarce. "He's
> that way because he lives at the bottom of a ditch and never gets any
> water," is an accepted explanation for even the most aberrant behav-
> ior in this valley (Crawford 1988:23–24).

The valley that Crawford evokes so well lies in northern New Mexico,
a region where Spanish settlers first planted their complex tradition of irrigation technology and water rights in 1598. Some of those ditches, then,
have been channeling water to people's fields for nearly 400 years. Today,
however, most of the landowners along the *acequia* (irrigation ditch) that
Crawford writes about work for wages in schools or government labora-
tories. Farming has become a part-time occupation, not a matter of life
and death.

About 450 miles to the southwest, on the other hand, those who
choose to stay on the land must live off the land, so the "arteries of ditches"
remain even more vital. In the *municipio* of Cucurpe, Sonora, Mexico, irrigated agriculture is the circulatory system of the local agrarian economy.
Without the web of earthen acequias, crops would wither, cattle would die,
and the connective tissue of soil, animals, and argument between neighbors would dry up and decay. There would be no pueblo of Cucurpe, no

comunidad (corporate community) of peasant rancher-farmers, no rural society with roots that extend back to the establishment of a Jesuit mission in the 1640s. The canal systems of Cucurpe, like those of northern New Mexico, are maddeningly vulnerable—to flood and drought, algae and erosion, seepage and evapotranspiration. But they have kept the flood-plain of the San Miguel River in cultivation for at least the last 350 years. Controlled by farmers who have organized themselves into local irriga-tion communities known as *comunes de agua,* they are a form of common property management dating back to Roman and Islamic Spain. Most of the time, members of the comunes elect their own leaders, manage their own water-control systems, and sanction water users who violate opera-tional rules independently of other political organizations such as the municipio, the corporate community, or state and federal agencies of the Mexican government.

Those tasks are not always carried out efficiently or without fric-tion, and the comunes have not yet extended their control over shal-low subsurface aquifers increasingly tapped by groundwater pumping. Given the Mexican government's increasing commitment to the privatiza-tion of agrarian resources, groundwater pumping may eventually threaten the survival of the irrigation communities, weakening or destroying the society of peasant rancher-farmers that depends on them. The physical and social challenges faced by Cucurpe's comunes de agua therefore re-flect challenges faced by small-scale irrigation systems across the world— challenges aggravated by the persistent failure of governments to recog-nize the enduring strengths of smallholder agriculture, even in the late twentieth century.

The Municipio of Cucurpe

The municipio (roughly analogous to a U.S. county) of Cucurpe is located in north-central Sonora, less than 100 miles south of the international bor-der (figure 2.1). It encompasses 1,789 square kilometers of rugged basin and range country dissected by the San Miguel River and its tributaries, the Río Dolores and the Río Saracachi. Cucurpe's economy is overwhelm-ingly agrarian, with about 82 percent of the work force making their living as private ranchers, peasant rancher-farmers, farm laborers, or cowboys. Mining, once important, employs only 10 to 15 percent of the working population. Many Cucurpeños (inhabitants of Cucurpe) migrate to the

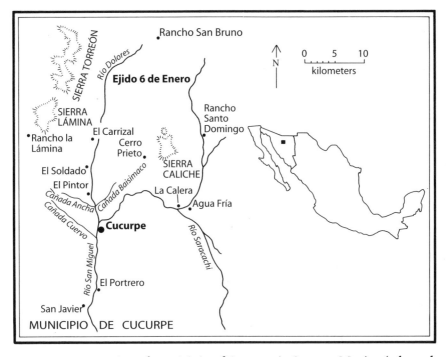

Figure 2.1   Location of municipio of Cucurpe in Sonora, Mexico (adapted from Sheridan 1988).

cities of Mexico or the United States to look for work, but those who stay behind remain tied to a world of cattle, horses, and fields (Sheridan 1988).

The fields constitute less than 1 percent of the municipio's total area. The rest of Cucurpe is too steep or dry to cultivate, so livestock raising, particularly cattle ranching, dominates the countryside. Nonetheless, agriculture is an essential component of Cucurpe's mixed agropastoralist economy. Much of the municipio falls into the Arizona Upland subdivision of the Sonoran Desert. Most of that mesquite grassland or desert scrub has been severely overgrazed. As a result, stockmen often have to provide supplementary fodder for their animals to keep them alive. The fodder comes from *milpas* (irrigated fields) along the narrow floodplain of the San Miguel, or from *temporales* (rainfall-dependent fields) along normally dry arroyos in a few major side drainages such as Cañada Ancha, Cañada Baisimaco, and Cañada Cuervo. With average annual precipitation ranging from 400 to 450 millimeters (15.76 to 17.73 inches) across the municipio, true dry farming is impossible.

Irrigated land, in particular, is in short supply. There are approximately 636 hectares of milpa in Cucurpe. Temporales cover nearly as much ground (506 hectares), but rainfall-dependent agriculture is a notoriously chancy endeavor in the Sonoran Desert. The rains may not come, or they may come too early or too late. Oftentimes, not enough rain falls to bring a crop to maturity, so the farmer gets little more than forage for his animals. Most Cucurpeños, then, consider temporal agriculture "*una lucha,*" a battle. Consequently, they devote about ten times more labor per hectare to their irrigated fields.

Nearly all those fields are located along the San Miguel, the Dolores, or the Saracachi. Like most "rivers" in the Sonoran Desert, however, the San Miguel and its tributaries are intermittent streams; they never flow the length of their channels except during floods. The rest of the time, stretches of surface water alternate with dry expanses of alluvial sand. Surface flow is fed by *nacimientos* (springs) in the riverbed — shallow currents of water that meander past man-made rows of cottonwood and willow, nourishing beds of watercress and attracting kingfishers, great blue herons, and many other species of birds. Cucurpeños throw diversion dams (*represas*) across those streams wherever the floodplain widens enough to trap and hold substantial deposits of alluvial soil. The represas, in turn, direct the flow into the earthen ditches that bring the water to the fields of the Cucurpeños themselves.

A few statistics reveal the importance of these simple, gravity-flow acequia systems. In 1980, about 1,200 people lived in the municipio. Those people were organized into approximately 260 households, at least 105 of which controlled irrigated fields. And of those 105 households, 97 (92 percent) relied, at least in part, upon the surface water of the San Miguel. Pump-powered wells have been a part of Cucurpe life since the 1940s, but the vast majority of farmers continue to wrestle with the river for much of the water they need.

The Physical System

Within the municipio, there are at least eighteen different acequias. They range in length from less than 100 meters to several kilometers, but most are simple and often ramshackle affairs. The gravity-flow systems begin with the represas, which are nothing more than mounds of sand and brush piled behind stout branches pounded into the channel of the river. All of the building material is local—alluvium and riparian vegetation such as seep willow (*Baccharis salicifolia*), desert broom (*Baccharis sarathroides*), burrobush (*Hymenoclea monogyra*), and tree tobacco (*Nicotiana glauca*). The only tools required are shovels, axes, machetes, and an occasional mule-drawn *escrepa* (scraper). The represas cannot survive even moderate floods, but they are easy to repair or replace.

The ditches are more permanent but not much more elaborate. Most are earthen channels—0.5 to 2 meters wide and 1 to 1.5 meters deep—carved into the banks of the floodplain. Along stretches where the floodplain narrows, the acequias usually hug canyon walls, where they occasionally are protected by living fencerows of cottonwood and willow trees. Those stretches, known locally as burros, are the most vulnerable to floods, and often get washed out along with the represas. To prevent that from happening, some enterprising farmers have constructed masonry sections that cling to the cliffs. Just north of the pueblo of Cucurpe, where the serpentine Río Saracachi curves toward its juncture with the Río Dolores, water users even blasted a tunnel through a rock outcrop. But such examples of labor intensification and technological sophistication are rare. Most acequias, like most represas, are pick-and-shovel creations. Their value, of course, is their simplicity. The acequias require no special machinery or training, and they cost very little to construct or repair. Local water users build and maintain them without interference from government officials and without resort to bankers, engineers, or heavy equipment operators. They are both autonomous and autochthonous, well-adapted in many respects to peasant subsistence agriculture and peasant social organization with its high value upon self-sufficiency and independence.

Nevertheless, autonomy has its price. As durable as the acequia systems are, they have severe limitations. The most basic, of course, is their absolute dependence upon surface flow, which fluctuates from season to season and year to year. The systems cannot impound water during periods of heavy flow in order to store it for periods when flow diminishes. The problem is particularly acute in late spring and early summer, the driest time of the year in the Sonoran Desert. Many springs slow to a trickle. Other nacimientos dry up completely. As a result, essential spring food crops such as corn, beans, and potatoes, planted in March, often do not receive the water they need during critical stages of their growth. During those months, farmers complain that their plants are "pleading" or "begging" for water, but the acequia systems cannot answer those pleas. The acequias are slaves to their rivers; they take what the rivers have to offer rather than transform the rivers themselves. That leaves them at the mercy of drought, one of the twin scourges of arid-lands agriculture.

The other scourge, paradoxically, is flood. In Cucurpe, more than two-thirds of the mean annual rainfall drenches the countryside in July, August, and September. These convectional thundershowers, known locally as *las aguas,* often send flash floods surging down the watershed, washing away represas and exposed sections of the canals. Because of the disruptions, some farmers do not plant at this time. Consequently, the acequias often fail Cucurpe farmers just when those farmers need them the most.

The Social System

The rudimentary nature of Cucurpe's acequia systems is not atypical. As Michael Meyer (1984:41) points out in his history of water control in the Hispanic Southwest, "The irrigation systems of most northern communities were scarcely models of hydraulic engineering. Skilled *acequeros* (ditch specialists) were scarce, surveying techniques crude, and communities had to rely on the talent that was available." In other words, farmers from Sonora to Texas suffered from many of the same limitations that aggravate Cucurpeños today. "The only predictable problem, but one without a predictable solution," Meyer (1984:45) concludes, "was that water scarcity would occur and would bring with it social polarization and confrontation. The variety of water conflicts was limited only by human ingenuity, but the conflicts themselves, as well as their solutions, fashioned some of the most important configurations of southwestern society."

One of those configurations is the *común de agua,* or local irrigation community. In Cucurpe, there are three formally organized levels of government in the municipio. The most encompassing is the municipio itself, which is governed by a *presidente* (president) elected every three years. The municipal government performs a variety of functions somewhat analogous to those of a county government in the United States.

The second level is the peasant corporate community, of which there are three in the municipio—the Comunidad de Cucurpe, the Comunidad de San Javier, and the Ejido 6 de Enero.[1] These federally chartered peasant organizations control approximately 21 percent of all land in the municipio—land that, according to Mexican agrarian reform law, is held in trust for the exclusive use of the members of the organizations alone. The corporate communities therefore control two of the basic resources—arable land and grazing land—upon which Cucurpe peasant rancher-farmers depend (Sheridan 1988).

Surface irrigation water, on the other hand, is controlled by local water users. As noted earlier, there are at least eighteen acequias in the municipio. Twelve of these ditches irrigate the fields of only a handful of farmers, so they are not subject to formal organization. Six of the canals, on the other hand, serve numerous farmers. Consequently, they fall under the jurisdiction of three different irrigation communities: the Común de Agua El Pintor—Cañada Ancha along the Río Dolores, the Común de Agua Saguarito—Hornitos at the junction of the Saracachi and the Dolores near the pueblo of Cucurpe, and the Común de Agua El Molino—Cerro Blanco south of Cucurpe along the San Miguel. In 1978, El Molino—Cerro Blanco had fifteen members farming about 35 to 40 hectares; El Pintor—Cañada Ancha had twenty-four members cultivating 175 to 200 hectares; and Saguarito—Hornitos had forty-three members with 120 to 140 hectares under irrigation.[2] Those eighty-two water-users constituted 78 percent of the 105 farmers with milpas in Cucurpe, and their total acreage represented approximately 50 to 60 percent of all irrigated land in the municipio.

Interestingly enough, all three of the comunes de agua lie within the boundaries of the Comunidad de Cucurpe, the largest and most powerful corporate community in the municipio. Nonetheless, the comunes, not the comunidad, make decisions about the water that flows down their canals. Each controls two major ditches, or *acequia madres,* on either side of the floodplain. Each is autonomous, responsible only to its members and to the presidente of the municipio. The comunidad exercises jurisdiction over the distribution of land, but surface irrigation water is allocated by the comunes according to a traditional system of water rights that has been passed down from one generation of farmers to another.

No one in Cucurpe knows how old these particular organizations are. The tiny municipal archive contains a scrap of paper noting that fourteen landowners were drawing water from Acequia Saguarito in 1937, but the irrigation communities undoubtedly date from a much earlier period. For example, a squabble over water rights in 1723 reveals the presence of a común in the pueblo that claimed jurisdiction over land and water along the Dolores and San Miguel rivers (Sheridan 1988:12–13). The documentary evidence is too fragmentary to determine precisely how this entity functioned. Because the modern comunes are similar to irrigation communities across Hispanic America, however, it is safe to assume that they distribute irrigation water in much the same fashion as their colonial predecessors (Hutchins 1928; Simmons 1972; Meyer 1984; Crawford 1988).

In theory, the associations are nearly as simple as the ditches they maintain. First and foremost, the comunes comprise only farmers who cultivate land along their acequias, not all members of the comunidad. Among peasant rancher-farmers in Cucurpe, household autonomy is highly valued. Most Cucurpeños view corporate tenure as a necessary evil, not a desired goal. People join together to control resources they cannot control on their own, such as grazing land or the surface flow of rivers, but they do so reluctantly. Whenever possible, their cooperative ventures are limited to specific tasks and specific resources. I call this deeply embedded logic of resource control the principle of the bare minimum. Irrigation communities control sources of surface water within comunidad boundaries, yet those associations, not the comunidad, regulate irrigation. Comunidad members who do not irrigate their fields from canals controlled by a particular común de agua have no say in the affairs of that irrigation community. Control has not been ceded to the higher level because the local comunes are able to perform the labor, make the investments, and regulate the water on their own.

The amount of water farmers receive roughly correlates with the amount of land they possess, but the correlation is seldom precise. Shares —and responsibilities—are determined by the traditional *hijuela* system, a term of venerable inexactitude. An hijuela can be anything from a small piece of cloth to a feeder canal off a main ditch. According to custom, one hijuela consists of ten to twelve *tareas*—a word that, literally translated, means the amount of work that can be accomplished during a certain period of time. In Cucurpe, the continued use of tarea harks back to an era when a unit of land was measured by how long it took to plow or sow or irrigate it, not by any absolute areal survey. Both hijuela and tarea, then, compound one another's vagueness.

The result is a system of water distribution with a considerable amount of inconsistency. During the general cleanings (*limpias generales*) that take place in February and August, each water-user must contribute a day's work (*jornal*) for each hijuela he or she possesses. One Cucurpeño complained that he was assessed the same amount of work—one jornal—as his brothers, even though he only irrigated five tareas of land while his brothers watered nine. It is easy to imagine the slow accretion of rights — and the endless negotiation of responsibilities — such a system represents.

The *Juez de Agua*

The individual who has to keep this Byzantine system working is the water judge (*juez de agua* or *comisionado de agua*). Every January, the members of a común meet to elect the juez de agua and his *suplente* (substitute). The water judge is always a fellow member with a vested interest in the común. Not surprisingly, in the Común de Agua Saguarito — Hornitos, which provides most of the examples that follow, two farmers whose fields lay at or near the end of the two acequia madres were often chosen to serve. They, of course, had more reason than most to see that the distribution of water was carried out in an equitable and efficient fashion. If they did not, their crops would shrivel while upstream water users hogged the flow.

One of the major tasks of the juez de agua is to organize the work parties that clean and repair the acequia systems. In February 1981, the limpia general of Saguarito — Hornitos lasted four days. The first day, work began at eight A.M. and ended at five P.M. with a break for lunch. Twenty-nine men — some of them water users, others *jornaleros* (day laborers) hired by water users — first replaced the diversion dam channeling water into Acequia Saguarito, the main canal on the west bank of the San Miguel River. Then they slowly moved down the ditch with their shovels and their machetes, removing debris, scraping away (*deslamando*) the algae that clogged the channel, trimming or chopping down seep willow, mesquite, or acacia overhanging the banks. Work progressed rapidly, but there were plenty of breaks for cigarettes and talk. The day was characterized by good-natured, easy-going industry, and Beto, the new juez de agua, was clearly pleased with his troops.

But that was during the winter, when flow in the San Miguel is most reliable. The most serious problems begin in late spring. In April 1981, within two months of the February general cleaning, Beto had to organize another work party to scour a stretch of Acequia Saguarito because algae (*lama*) was once again choking off flow. And when the summer rains began, members of the común spent eleven different days in August alone replacing the diversion dams or rebuilding stretches of Saguarito and Hornitos, the acequia madre on the east bank of the San Miguel.

Such emergency work groups are much harder to marshal than the limpias generales. Because floods and other natural insults to the system cannot be predicted, water-users or their jornaleros have to be rounded up after the fact. Some are busy. Others are not at home. The water judge has the power to fine those who do not show up, but many prefer to pay the money rather than invest the labor. As a result, the juez de agua spends more time bringing together fewer people, and the ones he does manage to corral often grumble about doing more than their fair share.

Nevertheless, maintaining the acequia systems generates far less conflict than distributing the water within their banks. At that same work party in early April, Beto was already worrying about the hot weather and the low flow. Water users irrigate in the order (*por turno*) in which their fields lie along the ditches. Because of the scarcity of water, Beto had ordered them to irrigate "*dia y noche*" ("day and night") when their turns came. That meant he had to make sure they were irrigating twenty-four hours a day and not falling asleep and letting the water go free. One night he even caught the farmer who had been water judge the year before snoozing beneath a tree. "*Levantate, muchacho!*" ("Get up, boy!"), he told his groggy neighbor, who was coming off a three-day drunk. In Cucurpe, where most members of the police force and other municipal agencies are temporary, an official one year can be a culprit the next.

It also meant that Beto had to watch out for unscrupulous water users who sneaked out at night and breached the acequias to water their own fields. When he confronted one notorious abuser, the man argued that gophers (*topos*) had caused the bank of the ditch to collapse. According to Beto, late spring was a time when the water judge had to exercise constant vigilance, when "the water cannot be left alone."[3]

By the middle of April, however, there was barely enough water in the river to steal. Flow in both Saguarito and Hornitos could not be sustained, so Beto had to divert all the water into one canal and then the other for five- and later eight-day periods. He also had to organize another work party to extend the intake canals upriver to capture more flow. At times like that, most water judges vow never to accept the job again. They are besieged by desperate farmers who beg for a few hours of "the tail end of water" (*la cola de agua*) to keep their bean plants from dying in the intense sun. But if the water judges grant one farmer's request, they face the wrath of others. By June, tempers are flaring as high as the summer heat.

To make matters worse, the tension between water judge and water users is aggravated by the fact that the juez de agua's annual fee—one *fanega* (about 1.5 bushels) or *saco* (sack weighing 75 kilograms) of wheat or its monetary equivalent per hijuela—is due that month. By the middle of June, Beto had received payment from only two members of the común. He undoubtedly would have agreed with his predecessor, who said, "The comuneros are very devious. They all want their water rights but they don't comply with their obligations. They're coyotes." [4]

With luck, the summer rains arrive by San Juan's Day (June 24) and break the tension with violent thunderstorms. In 1981, however, the rains began in late June but quit a month later. By the end of August, the drought was critical, and farmers were struggling to keep their corn, beans, and chile alive. Not surprisingly, their desperation often turned into hostility toward the water judge. Eleven days earlier, members of the común had repaired the diversion dams and the *canoa* (wooden flume) after the last flood. Nevertheless, an upstream water user, entitled to four or five days of water, was still dominating the flow. The same man who had stolen water at night and blamed it on gophers now damned the juez de agua. A devastating raconteur, he mimicked the water judge's laugh and spat, "This man (the water judge) is such an idiot! There's no vigilance! This fucking old man [the upstream water user] lets the water flow free and the water judge doesn't punish him. These are very irresponsible people." [5]

Other water users echoed his words. During a *velorio* (all-night vigil) to the Santo Niño de Atocha at a ranch south of Cucurpe, a group of farmers smoked and talked outside as the women prayed in the ranch house and the stars burned overhead. [6] They, too, noted that their fellow member—"*el pinche perro viejo*" ("the fucking old dog")—had been irrigating for nearly two weeks while their crops withered. They also complained about the juez de agua, agreeing that he wasn't doing his job. "When there's great necessity, people ought to irrigate day and night, as the regulations state," one proclaimed. "But the people have no conscience." [7] Moreover, food crops should be irrigated before fodder crops. Alfalfa could survive another three or four days, but bean plants were more vulnerable, especially when they were *en ejote* (in pod), as they were now. "*Pero la gente es muy perra*," he concluded. "People are such dogs."

Pumps and Pipe Dreams

Such derogatory comments were not uncommon, especially during times of stress. Several people stated, only half in jest, that the reason why the thunderstorms that were soaking the rest of the region bypassed Cucurpe was because "the people here are very bad, very bad."[8] When I jokingly suggested to an old friend that the Cucurpeños should pray for rain, he shook his head. "No, Tomás. They should unite. They should act as brothers."[9] He knew as well as I that such a change in attitude was unlikely.

The cynicism of the Cucurpeños reveals the social as well as physical limitations of the comunes de agua. The comunes function, but not without friction. Surface water may be a corporate resource, but its distribution is a constant source of conflict among water-users during the drier months. Moreover, the members of Saguarito—Hornitos had trouble mustering the organizational energy to improve their acequia system, even in relatively minor ways. In the late 1940s, water users hammered and blasted a tunnel through solid rock so Acequia Hornitos did not have to skirt an outcrop in the floodplain of the Río Saracachi, where it was washed out by every flood. Dynamite had to be purchased. Labor had to be mobilized. Planting along the canal was even suspended while the work progressed. In the early 1980s, however, the water users could not even replace the porous wooden flume that transported the water of Acequia Saguarito across the Río Dolores. Consequently, considerable amounts of water from Acequia Saguarito spilled into the sand of the Río Dolores instead of onto their fields.

But cynicism and conflict were not the only responses of the Cucurpeños to water scarcity. In meetings, on street corners, and in their fields, farmers spent hours discussing ways to get more water to their crops. One idea was to construct a concrete *cortina*, or retaining wall, from the surface of the Río Saracachi down to bedrock in order to force subterranean water to the surface. But the solution most Cucurpeños preferred was to drill pump-powered wells. They argued about where the wells should be placed. They debated the merits of deep, perforated wells (*pozos perforados*) versus shallow, open ones (*pozos de luz*). They even dreamed about drilling wells in the floodplain and then pumping water through tubes up and over the mesas to irrigate two fertile side canyons of the watershed. Like farmers all over arid North America, Cucurpeños lusted after a technological fix.

Pump-powered wells (*pozos con bombas*) have been a part of Cucurpe life for nearly half a century. The first one was drilled along the Río Dolores in the late 1930s or early 1940s. Others soon followed, largely along the Dolores where a few relatively wealthy farmers controlled considerable land. By 1980–1981, there were at least twenty-five pump-powered irrigation wells in the municipio, most of them located along the floodplain of the rivers themselves. Nineteen were open and relatively shallow, drawing water to the surface with gasoline-fueled centrifuge pumps. Along the Dolores, however, five farmers operated six perforated wells powered by diesel turbine pumps. Those wells were six to ten inches in diameter and pumped six to eight inches of water. All were in the floodplain of the Dolores or in a side canyon called Cañada Ancha. All tapped the same shallow, alluvial aquifer that fed the springs of the Río Dolores.

Traditional Water Control and Technology: A Disjunction

Significantly, four of those five well-owning farmers were bitter enemies of the comunidad of Cucurpe. Their land lay within comunidad boundaries, yet ever since the federal government recognized the comunidad as a legal entity in 1976, they had battled the *comuneros* (members of the comunidad) in both state and federal courts. The comunidad eventually won those suits, but the government, pressured by the landowners and their powerful friends, took no action. So in 1982, a group of determined comuneros occupied the fields of the four *quejosos* (complainers), three of whom lived outside the municipio. The government told the comuneros to get off the land but the comuneros refused. Eight years later, the land remained in their hands (Sheridan 1988:179–184).

Before that happened, however, their wells provided a stark contrast between corporate and private tenure. In Cucurpe, surface water in the rivers is held in trust by communities of water users. Water pumped from the ground, on the other hand, belongs to the owner of the well. Wells are registered by the federal Department of Agriculture and Water Resources (SARH; Secretaría de Agricultura y Recursos Hidráulicos), and it is now difficult and expensive to secure permission to drill a new one along the San Miguel. But SARH has not yet metered the wells in order to monitor or control pumping. Consequently, well owners are able to pump as much water as they can afford; the only constraint is fuel costs, not the needs or rights of neighbors. During the early 1980s, private wells often chugged and rumbled while flow in the rivers declined. Listening to those pumps one hot August afternoon, a former juez de agua compared his crops with the crops of two comuneros who sharecropped the fields of a well owner who lived in Magdalena. Their chile plants were dark green and lustrous and their corn had heavy ears. His beans, in contrast, were parched and struggling to survive. "How envious it makes me feel to see those fields!" he exclaimed. "It takes away your desire to work." [10] The only major difference between his fields and theirs, of course, was the availability of water.

The distinction between surface and subsurface water is an old one, rooted in Spanish law. Michael Meyer (1984) points out that in New Spain, the colonial viceroyalty encompassing Mexico and the southwestern United States, irrigation water, like land and mineral wealth, was part of the royal patrimony. Unless the king or his officials granted it to a community or a private individual, it belonged to the Spanish crown. Moreover, water rights were legally separate from land tenure. A grant of arable land along a river did not entitle the owner to water his or her crops from its flow. The two resources—land and river water—were distinct and had to be acquired as such.

Groundwater, in contrast, belonged to the owner of the land where it bubbled to the surface. According to Meyer (1984:20):

> The only automatic alienation of water with a land grant was for water that originated on the piece of land. A spring or a well became the property of the owner of the land where it rose, a tradition deeply engrained in medieval Spanish water law. The distinct ownership of surface and subsurface water is not easy to explain. Water originating from rain was considered common property, but knowledge of aquifers was very rudimentary. Maybe the water in springs and wells came from subterranean sources; maybe it had just always been there; the supply certainly seemed limitless. There was little or no appreciation that underground water also originated from precipitation, or that depleting an underground reserve on a given piece of property could have a direct impact on the water supply of a neighbor. Given this imperfect understanding, a person could pump water from a well or channel spring water to his fields without special permission.

Prior to the installation of pump-powered wells, the distinction probably had little impact upon Cucurpe irrigation practices. Subsurface water was largely inaccessible. All farmers, regardless of their wealth, relied upon the intermittent flow of the rivers. With the advent of this technological innovation, however, traditional methods of water control faced a potentially severe, even fatal challenge. Nearly all the pump-powered wells were drawing water from the same aquifers that fed the springs in the riverbeds, yet the comunes de agua had no jurisdiction over the wells that caused flow in their acequias to decline. To rub salt in the wound, the owners of the largest wells even sold water to their less fortunate neighbors. In 1980–1981, the going rate was seventy pesos ($2.80) an hour, a considerable expense.[11] This was particularly galling to many comuneros, who were forced to buy water from absentee landowners fighting the comunidad. Their resentment undoubtedly fueled their move to occupy the land a year later.

The Future of Water Control in Cucurpe

The conflict between well owners and the comunes de agua is not yet critical, but it may become so in the years ahead, especially if the privatization of resources within comunidades and *ejidos* instituted during the presidential administration of Carlos Salinas de Gortari (1988–1994) continues. Agricultural intensification is impossible in Cucurpe without increasing the supply of irrigation water. At present, drought-sensitive plants like beans often suffer, while many lucrative crops such as chile, alfalfa, and vegetables sometimes cannot be grown without a well. The comunes have a hard time providing enough water for land under cultivation now. As long as they rely solely upon their gravity-flow acequia systems, expansion is out of the question. Consequently, Cucurpeños have largely ignored the comunes in their attempts to bring more water to their fields.

Three basic strategies have been pursued. The first is the sinking of private wells. Many farmers have done so, including many comuneros, but their wells are usually shallow and powered by gasoline pumps that are expensive to run, break down easily, and provide relatively little water. They are, at best, auxiliary sources of irrigation.

The second strategy is government intervention. In the late 1970s, SARH spent hundreds of thousands of pesos drilling deep, perforated wells in three major side canyons of the San Miguel. Those turbine-powered wells—several as deep as 500 feet—were supposed to deliver 10 to 12 inches of water in diameter, thereby opening up hundreds of hectares of potentially arable land. But the wells were colossal failures, barely pumping enough water to fill a few cattle tanks. In Cañada Ancha, technicians even abandoned their drilling rig. It served as a mute but massive reminder of technical ineptitude and governmental failure.

The third strategy is social reorganization. In the early 1980s, the government proposed a plan to funnel credit and technological assistance to small groups of farmers within the comunidad of Cucurpe who cultivated land next to one another. When first presented, these "collective groups" (*grupos colectivos*) were supposed to tear down their fences and farm their land in common. If they agreed to do so, the government was going to underwrite loans from the Banco Rural, the federally funded rural development bank, to drill wells, purchase agricultural equipment, and buy fertilizer, insecticides, and seeds.

The plan must have seemed like a good one in Mexico City, but it met unyielding resistance from the Cucurpeños. Deeply attached to their fields, they were not about to surrender their independence to either their neighbors or the Banco Rural. "We are going to become workers for the bank," one concluded after listening to the bureaucrats during a special meeting of the comunidad. Eventually, the bureaucrats got the message, and the proposed experiment in small-scale collective farming was abandoned (Sheridan 1988:186–88).

Most government officials viewed this resistance as yet another example of peasant conservatism and intransigence. Once the specter of collectivity was banished, however, many Cucurpeños expressed keen interest in government assistance to achieve more limited goals, such as the formation of pig-raising or bee-keeping cooperatives. In 1981, three different groups of comuneros even organized grupos colectivos. All were composed of farmers with temporales in the side canyons of the San Miguel. All sought loans to drill pump-powered irrigation wells.

Unfortunately, the economic crisis of the 1980s prevented the government from following through on its promises. The collective groups, like so many other government schemes, collapsed in a morass of corruption, devaluation, and retrenchment. Nonetheless, one group of five farmers, stimulated by the government plan, pursued matters on their own. Purchasing six hectares of temporal land in Cañada Baisimaco north of Cucurpe, they drilled a 70-foot well with a 6-inch diameter pipe and converted the rainfall-dependent fields into irrigated milpas. It was one of the few successful examples of cooperative agricultural intensification in Cucurpe during the past fifteen years.

It also revealed that fundamental premise of resource control in Cucurpe: the principle of the bare minimum. When left to their own devices, Cucurpeños sought the most parsimonious solutions to their problems. If they could afford to drill a well themselves, they did so. If not, they banded together in the smallest group possible. Centuries ago, farmers who needed water from the same stretch of the San Miguel River formed comunes de agua. Today, farmers with the means to do so tap subsurface reserves.

Unfortunately, the result is a patchwork quilt of wells and ditches that extract water from the same aquifers without necessarily benefiting the community of water users as a whole. Technology has outstripped traditional methods of water control. The possibility of conflict increases faster than the reality of cooperation, particularly as the Mexican government abandons its institutional commitment to common property management and encourages privatization. Consequently, economic inequality and political polarization may deepen as those who are able to pump water from the ground diminish the livelihoods of those who cannot. Continuing to view subsurface water as a private resource may lead to even more crippling water scarcity for those who depend upon surface flow.

In order to prevent this from happening, further centralization of water control may be required. The critical factor in this process, however, is at what level centralization should take place. In Cucurpe, the impact of pumping upon surface flow in the San Miguel drainage needs to be ascertained — a task that probably can only be carried out by hydrologists supported by the federal government. Once that has been done, the artificial distinction between surface and subsurface water rights should be modified or eliminated. If they are to remain in private hands, existing wells should be strictly regulated. Better yet, those that pump water from the same aquifers that supply the comunes de agua should be incorporated into those organizations. In San Ignacio, a farming community about 40 miles northwest of Cucurpe, one such irrigation community drilled a well that serves as a supplementary source of irrigation water. To pay for the well, members were assessed according to the same ratio as their contributions to the communal labor parties. As a result, members of that común have enough water to raise a wide variety of fruits, vegetables, spices, and even flowers for the regional market — crops the Cucurpeños simply do not have enough water to grow.[12]

Unless such appropriate centralization takes place, however, the disjunction between technology and traditional methods of water control—and between private and corporate tenure—will continue to widen. Despite their limitations, the comunes de agua have allowed peasant farmers to survive in Cucurpe for nearly four hundred years—an historical reality the Mexican government should not ignore in its rush to privatize the nation's agrarian economy. As Jonathan Mabry points out in the first chapter of this volume, 85 percent of the world's irrigated cropland continues to be watered by small-scale gravity-flow systems managed by local groups of farmers. Battered by devaluation and plagued by high rates of unemployment and underemployment, the service and industrial sectors of Mexico's urban economy cannot absorb the additional millions of peasants who will be forced to leave the countryside if the enduring forms of common property management protected by Mexico's ejidos and comunidades are weakened or destroyed. With a limited amount of government intervention, both legal and financial, pump-powered wells could be placed under the jurisdiction of Cucurpe's comunes de agua, thereby insuring all local water-users their fair share of irrigation water. Otherwise, the anarchy of private tenure within a corporate framework will erode the foundations of a rural society that provides a decent living for hundreds of people, just as other forms of privatization and so-called structural adjustment are threatening the livelihoods of millions of peasants across the world. In the words of Stanley Crawford (1990:265),

> Everyone has been startled at least once by the miraculous imagery of the saying that water can be made to flow uphill toward money. It is perhaps part of the genius of the image that for the moment at least it seems to settle the arguments that bring it into play. Everything has its price. We live in a world of buying and selling. We all know that. What else can we say?
>
> Yet we also live in a world of other kinds of value, and sooner or later we will need to ask whether it is good for anything but money that water can be made to flow uphill, whether it is bad for streams, rivers, traditions, communities, and even individuals.

Notes

1. In 1976, the Mexican government reconfirmed the existence of the comunidades of Cucurpe and San Javier, and created the Ejido 6 de Enero. Cucurpe was granted 21,050

hectares, San Javier 9,867 hectares, and 6 de Enero 8,000 hectares, for a total of 38,917 hectares of the 189,001 hectares in the entire municipio of Cucurpe. The rest of the land belongs to private ranchers. These organizations, similar to peasant corporate communities around the world, hold land in trust for their members (Wolf 1955, 1957; Sheridan 1988). Only members are legally entitled to cultivate arable land, pasture their livestock, and gather firewood and other wild plant resources within their boundaries. Like comunidades and ejidos all over Mexico, they elect their own officers, hold monthly meetings, and assess dues in accordance with Mexican agrarian reform law.

2. The relatively large amount of acreage irrigated by Común de Agua El Pintor — Cañada Ancha included the fields of the four landowners fighting the Comunidad de Cucurpe.

3. *No se puede dejar la agua solo.*

4. *Los comuneros son muy mañosos. Todos quieren sus derechos de agua pero no sirven. Son coyotes.*

5. *Este hombre tan idiota! No hay vigilancia! Este pinche viejo deja el agua solo y no le castiga. Son gente muy irresponsable.*

6. The Santo Niño de Atocha is a manifestation of the Child Jesus who appeared in Atocha, Spain, during its occupation by the Moors. According to legend, the Moors allowed no one but small children to visit Christians in prison there. One day a child carrying a basket of bread and a staff with a gourd of water suspended from one end showed up at the gates. The child fed and gave water to all the prisoners, yet his basket and gourd remained full. In Cucurpe, at least one family appealed to the Santo Niño to end the drought.

7. *Cuando hay mucha necesidad, la gente debe regar dia y noche, como dice el acto. Pero la gente no tiene conciencia.*

8. *La gente acá son muy malos, muy malditos.*

9. *No, Tomás. Deben unirse. Deben hermanarse.*

10. *Como me da envidia a ver esas milpas! Se quita el ánimo a trabajar.*

11. In 1980–1981, the exchange rate was approximately twenty-five Mexican pesos per U.S. dollar.

12. San Ignacio, established as a Jesuit mission among the Upper Pima Indians in 1687, is located along the Río Magdalena. In the spring of 1990, one family of farmers who belong to the común de agua that drilled the supplementary well raised spinach, carrots, radishes, onions, garlic, cilantro (Mexican coriander), chamomile, Chinese cabbage, and peaches in one small field alone. Other farmers cultivate orchards of plums and quinces and fields of marigolds, which decorate graves on El Día de los Muertos (All-Souls' Day), in addition to wheat, barley, alfalfa, chile, corn, beans, and squash. Proximity to Magdalena, a regional center of about 30,000 people, and Highway 15, which leads to Nogales and the international border, are two reasons why truck farming flourishes in San Ignacio. Even with access to such markets, however, Cucurpe farmers could not grow many of those crops without additional sources of water.

References Cited

Crawford, Stanley

1988 *Mayordomo: Chronicle of an Acequia in Northern New Mexico.* New York: Doubleday.

1990 Dancing for Water. *Journal of the Southwest* 32(3):265–67.

Hutchins, Wells

1928 The Community Acequia: Its Origins and Development. *Southwestern Historical Quarterly* 3:261–84.

Meyer, Michael C.

1984 *Water in the Hispanic Southwest: A Social and Legal History, 1550–1850.* Tucson: University of Arizona Press.

Sheridan, Thomas E.

1988 *Where the Dove Calls: The Political Ecology of a Peasant Corporate Community in Northwestern Mexico.* Tucson: University of Arizona Press.

Simmons, Marc

1972 Spanish Irrigation Practices in New Mexico. *New Mexico Historical Review* 47(2):135–50.

Wolf, Eric

1955 Types of Latin American Peasantry. *American Anthropologist* 57:452–71.

1957 Closed Corporate Communities in Mesoamerica and Central Java. *Southwestern Journal of Anthropology* 13(1):1–18.

3

Dhasheeg Agriculture in the Jubba Valley, Somalia

Catherine Besteman

For at least the past 5,000 years, flood recession agriculture has been vital to the people living in Africa's river valleys. As Richards (1983:29) notes, flood recession regimes are "natural irrigation systems" that obviated the need to develop complex water control (irrigation) schemes by African communities. Despite the widespread dependence on flood recession in African agriculture, Scudder (1992) argues that "a striking feature of large-scale post-World War II river basin development projects is the extent to which the importance of annual flooding for local farmers, transient pastoralists, and fisherfolk is ignored by both researchers and developers alike." This ignorance of indigenous systems of water use may go a long way toward explaining the findings of a recent, detailed overview of development projects in African river basins: "Many of the projects that have been built have failed completely; many have been disappointing; some have been modestly successful for at least a few years; unfortunately there are none known to us that can be considered unqualified successes" (Bloch 1986:1).

With growing concern over the "crisis" in African agriculture (Barker 1984:12; Berry 1984, 1989; Commins, Lofchie, and Payne 1986; Richards 1983:2; Watts and Shenton 1984:173, 176) and a continuing use of "technocratic approaches" (Berry 1989:1) such as dams and irrigation projects to meet this crisis, a better understanding of how indigenous local communities have successfully used and managed their floodwater resources is warranted. This perspective calls for attention not only to land use strategies, but also to land tenure patterns—the ways in which access to flooded land is distributed, obtained, and maintained by community members.

Unfortunately, not only are detailed analyses of land use patterns in flood recession regimes noticeably lacking in the literature (a notable exception is Scudder's 1975 discussion of Gwembe Tonga) but also lacking are analyses of the social organization of flood recession agriculture (with the exception of Park 1988, 1992, and Linares 1981). In this chapter I address these issues as they apply to one of Africa's "major river systems" (Scudder 1992), the Jubba River Valley of southern Somalia.

A particularly important aspect of the social organization of agriculture is the potential for an enhancement of pre-existing stratification or the emergence of new forms of stratification with development. The "equity" issue in African agriculture has received increasing attention (see Downs and Reyna 1988; Reyna 1987; Cohen 1980; Okoth-Ogendo 1976; Watts 1983). Concerns about equity with development are justified for economic as well as social reasons. In some areas where land concentration resulted from agricultural policies intended to facilitate development, agricultural productivity on larger farms fell (Haugerud 1989; Cohen 1980; see also Netting 1993). Competition for land is often particularly intense in river valleys, due to the high value placed on irrigated (or flood recession) land. As Richards (1983:29) notes, " 'irrigated' land . . . is often keenly sought after and rights to it jealously guarded. Inequality in access to swamp land may in some circumstances be a better clue to socioeconomic stratification among the peasantry than size or availability of upland farms." Comparatively, equitability in the distribution of access to wetlands is one area of high variability. One purpose of this essay is to examine the issue of unequal access to and control of flood recession land among local Jubba valley farmers.

Drawing on data collected during twelve months of fieldwork in a village in the mid-valley, I discuss the importance of *dhasheeg* agriculture to subsistence, analyzing risk minimization strategies used by local farmers. My point is to demonstrate how flexibility is maintained and stratification stifled in the social organization of flood recession agriculture. I will conclude with some remarks on the nature of the conflict between local management of resources and state intervention in local resource management (in the form of new tenure laws, titling programs, development projects), and the place of irrigable land in contemporary war-torn Somalia.

Middle Jubba Ecology and Geography

The middle portion of the Jubba River valley (figure 3.1) is distinguished from the upper and lowest part of the valley by the prevalence of inland, low-lying depressions called dhasheegs, which are seasonally flooded by river channel overflows, rainfall, run-off, and underground water flow (percolation). The dhasheegs, rather than the riverbanks, are the focus of

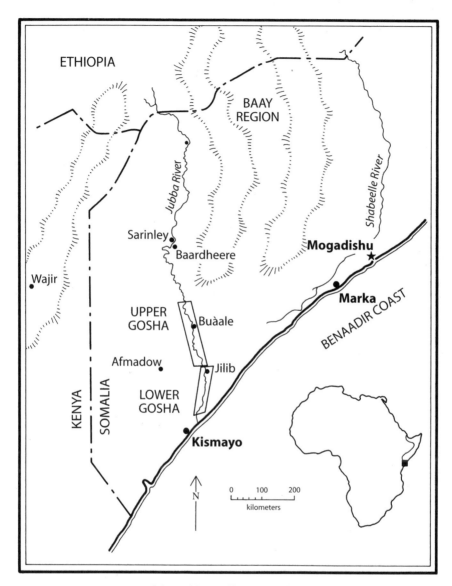

Figure 3.1 Location of the Jubba Valley, Somalia (adapted from Michelin map no. 154).

flood recession agriculture. Dhasheegs are characterized by heavy black soils (*caramaddow*) that hold water and retain moisture for long periods of time and can yield up to twice as much as surrounding rainfed croplands (called *doonk*).[1] When full, dhasheegs resemble shallow lakes, and are cultivated from the perimeter toward the center as the water evaporates. The majority of Jubba Valley farmers practice no form of mechanical irrigation, and are thus dependent upon rainfall and seasonal flooding to meet their water needs.

Floods and rains occur in a bimodal pattern, but are highly unpredictable. Statistics of flooding for the Jubba Valley show a "wildly erratic seasonal and annual pattern. Thirty-year records show a range in annual floods of 2,700 million cubic meters (mcm) (1980) to 10,200 mcm (1977), while measured daily discharges at Luuq can range from 2,000 cubic meters per second in October to less than 10 cumecs in March in some years" (Tillman 1988:3). Rainfall patterns fluctuate in a similar manner. Recent records from Jilib in the lower valley show a range in annual rainfall from 472 mm (1980) to 925 mm (1983) (FEWS records). The average yearly rainfall in the mid-valley region is 400 mm (GTZ 1984).

In an area characterized by such unpredictable water availability, dhasheegs provide the key to survival. Because dhasheeg soils are so fertile and retain moisture for long periods of time, farmers say "there are no droughts on dhasheeg land." Dhasheegs will always produce something, even with minimal amounts of rainfall, whereas doonk farms are only cultivated in seasons of good rainfall or extremely high flooding. Dhasheegs thus represent a type of intensive agriculture in an area characterized by shifting cultivation, as dhasheegs are permanently cultivated at least once and often twice a year.

Dhasheeg land is valued for other reasons as well. Labor inputs are lower as flooding checks the regeneration of bush, so no land preparation is necessary prior to planting. Weeds are less of a problem than on doonk land. Planting, weeding, and harvesting continuously follow the receding water line, so labor bottlenecks are avoided. The planting of dhasheegs does not begin until after the water from the rains or floods has started to recede, so harvesting may not begin until well into the dry season, a time when fresh produce is much welcomed. Because dhasheegs are basins completely cleared of forest and bush, watching for animal pests is much easier (hippos, warthogs, monkeys, and cattle and camel herds of pastoral nomads are a primary source of crop damage), and neighbors can share guarding.

Dhasheegs and the Settlement of the Valley

Dhasheegs were critical in the settlement of the Jubba Valley for agriculture. Linguistic evidence and oral tradition suggest that Somali pastoralists, expanding into the Horn of Africa from their place of origin in southwestern Ethiopia, crossed the valley by the thirteenth century, followed by Boran pastoralists, who dominated the area until the late eighteenth century.[2] In the nineteenth century, Darood Somali groups began their great series of southward migrations, pushing the Boran all the way back to the Tana by the end of the century. Somalis have since controlled the plains stretching to the east and west of the Jubba. In the course of their seasonal movements, these pastoralists used watering places on the river and in adjacent dhasheegs for their livestock. Generally, however, the lower half of the valley was avoided by pastoralists except in times of great need, as it was dense jungle infested with malaria-bearing mosquitoes and tsetse flies that carry tripanosomes, vectors of sleeping sickness that is lethal to livestock. One indication of the Somali opinion of the Jubba Valley is the contemporary term used to signify the lower half of the valley—"Gosha"—which means "jungle or unhealthy place."

There is no conclusive evidence—archaeological, documentary, or in oral tradition—to suggest that agriculture was practiced prior to the arrival in the mid-nineteenth century of the forefathers of the present day farmers.[3] The first sedentary settlers in the valley were fugitive slaves. A plantation mode of production, using slave labor brought from East Africa, was well established in the Shabelle River valley by the mid-nineteenth century.[4] Starting in the 1840s, slaves escaping from these plantations began arriving in small numbers in the Jubba River valley. The number of ex-slaves coming into the valley grew dramatically by the late 1800s (Menkhaus 1989), when a British expedition up the Jubba estimated the riverine population (collectively called the Gosha, after the area in which they live) at 40,000 (Dundas 1893). The uninhabited valley, with its dense forest cover and unhealthy environment, provided a safe refuge for the escaping slaves. By the turn of the century, the fugitive slave population had conquered the indigenous foraging Boni and emerged victorious from a series of armed skirmishes with surrounding Somali pastoralists (Menkhaus 1989).

From nineteenth-century travelers' accounts, oral tradition and colonial documents we have a good understanding of the early settlement pattern of the Gosha.[5] The earliest arrivals settled near a dhasheeg, obtained seeds through a reciprocal agreement with a neighboring Somali group, and began farming in the dhasheeg. The forest provided abundant food until returns from agriculture were enough to meet subsistence needs.[6] As migrants streamed into the valley over the next several decades, new villages were continually being formed. By the last decade of the nineteenth century, the settlement pattern was characterized by numerous small, dispersed, independent villages.[7] Dhasheegs were critical in determining the settlement pattern, and in the survival of the fledgling villages. All villages were settled near a dhasheeg (Menkhaus 1989; Besteman 1991), and "most of the cultivation was done in *desheks*" (Cassanelli 1989:223). An early Italian administrator writing on the settlement of the Gosha called dhasheeg agriculture "importantissimo" (Zoli 1927) and later colonial authorities through the 1950s wrote of its significant potential for agricultural development (see Conforti 1954).

Cassanelli (1989) provides evidence that village founders and their descendants (*gamas*) maintained a greater degree of control over land than newcomers (*majoro*). The frontier environment of the Gosha stifled the development of extreme social stratification, however, as villagers could pick up and continue north if local politics were not to their liking. There was thus an effective limit to the authority of the gamas, as there were no means of preventing disgruntled villagers from founding villages of their own. Villages were largely independent and autonomous entities, and were continually established throughout the Gosha, particularly in a northward direction. The expansion of the frontier continued into the mid-valley area through the first decades of the twentieth century. Significantly, very little settlement of new villages continued beyond the geographical area characterized by dhasheegs.

Dhasheeg Land Use and Tenure

A case study of Loc village can provide an example of the workings of dhasheeg land tenure and land use in a mid-valley village. The village of Loc is similar in settlement history, land use, and land tenure patterns to other

Gosha villages, and can be considered representative. Loc was founded around 1920 by fugitive slaves from the Shabelle River Valley and Baay region. The first settlers had lived in several other villages in the Gosha before finally settling near the large dhasheegs in the vicinity of Loc. The area was initially settled in four villages, each near a dhasheeg. In 1977 the government relocated the four villages to the present site of Loc. The area currently claimed by Loc covers about 751 hectares and includes part or all of the four nearby dhasheegs, representing about 10 percent of Loc's land base. The earliest arrivals farmed the most accessible, highest quality dhasheeg land. As later settlers arrived, they claimed and farmed land further into the dhasheeg, which was less accessible, or land on the edge of the dhasheegs. Although still good quality, these lands are less desirable than the land claimed first. The land further into the dhasheeg, due to its location nearer the center, can become waterlogged; conversely, land on the fringes of the dhasheeg may not receive enough water.

Initial acquisition of land was based on the premise that land belonged to the man who cleared it. As villages formed in the area, village leaders emerged (called *nabadoon,* which means "peace-bringer"). These men, who were usually the village founders, were responsible for giving land rights to newcomers. The position of nabadoon became more important as population and demand for land grew. By the 1960s, the newcomers to the Loc area were actually giving cash payments to local nabadoons in return for rights to a piece of land. In 1977, with the resettlement of the four villages into Loc, the nabadoons were replaced by a government-appointed village council. The village council, consisting of six members, had the same duties as the nabadoon, including the authority to grant land to those who request it. An individual desiring land usually identifies an area of unclaimed land that he would like, and then makes a formal request to the village council for the land. The council is then responsible for ascertaining that there are no competing claims. If there are none, then the petitioner is given the authority to claim the land as his own.

Land acquisition in dhasheegs was characterized first by direct claim, and later, as population grew, by request through the nabadoon. Most dhasheeg land was claimed by about forty years ago, and inheritance has become the primary means of acquisition. A small amount of dhasheeg land continues to be acquired each year by other means, through the village council, by purchase, or as a gift. Dhasheeg parcels acquired through the village council over the past twenty years have generally been the poor-

Table 3.1 Means of acquisition of dhasheeg land parcels in Loc Village, Somalia.

How acquired	Percentage of dhasheeg parcels
Inheritance	68.6
Village council	15.8
Purchase	7.8
Gift	7.8

Source: Author's field research 1987 to 1988.

est quality, the land still characterized as being of caramaddow soil, but located on the extreme periphery of the dhasheeg. Purchased dhasheeg land is uncommon, and has occurred only when a farmer is moving away from the area and chooses to sell his land rather than gift it. Receiving dhasheeg land as a gift is equally rare, and in Loc has taken the form of either very long term borrowing (borrowing in perpetuity) or donations to important religious leaders in the community. Table 3.1 shows how currently held dhasheeg parcels were acquired by Loc farmers.

Currently, although less than 10 percent of the farmland of Loc is dhasheeg, dhasheeg farms represent 28 percent of the total cultivated area, and 38 percent of the total number of cultivated parcels in Loc. These figures underscore the desirability of dhasheeg land for Loc farmers.

Evidence of Differentiation in Dhasheeg Holdings

Permanent access to dhasheeg land is virtually assured in Loc. Eighty-eight percent of Loc households hold dhasheeg land; those who do not hold land are either very new to the community or have chosen not to acquire dhasheeg land because their upland farms are extensive, dependable, and of good quality. Differences between households are found, however, in the quantity and quality of dhasheeg holdings.

Although there is a strong correlation between the amount of land held by a household and the size of that household, no such correlation exists in regard to the amount of dhasheeg land held. Whether or not one is descended from one of the earliest settlers is a much more important determinant of the amount of dhasheeg land held. The average amount of total dhasheeg land held by the descendants of the earliest settlers is 2.6

hectares per household.[8] The average amount held by everyone else is 1.7 hectares per household.

Differentiation is also seen in the quality of land. As discussed earlier, the earliest settlers claimed the best land. The best land in the dhasheeg is the land that is most accessible, not in the middle (which gets water-logged) and not on the periphery (which may not receive enough water and where the black dhasheeg soil begins to mix with "white" doonk soils, which do not retain moisture well). Later acquisitions, and especially the most recent acquisitions, are of poorer quality. So although dhasheeg land continues to be accessible by request to the village council, the best dhasheeg land has been claimed for generations by the descendants of the first settlers. In the most extreme case, the vast majority of one dhasheeg is owned by one farmer in the village. His father was the first settler in the area, laying claim to about 6 hectares of a dhasheeg, encompassing the best land. He left only one son, the present owner, who has maintained control over the entire holding.

Thus stratification of control over dhasheeg land is evident. Unequal access is seen in terms of quantity and quality of dhasheeg land held, with descendants of early settlers having better access due to inheritance. The stratification is mitigated by several important factors affecting access to dhasheeg land. First, the size of inherited parcels is declining over time. Dhasheeg land is the type of land most likely to be divided at inheritance. With continued high population growth, partible inheritance, and no out-migration (or very little), the tendency to fragment parcels will certainly continue. The size of newly acquired dhasheeg parcels (through the village council) has increased over time, although the land is of poorer quality. Within the last ten years, and for the first time in Loc history, the average size of dhasheeg parcels inherited has been smaller than the average size of noninherited dhasheeg parcels. This tendency will facilitate an evening out of the amount (although not necessarily the quality) of dhasheeg land held per household, at least over the next decade or so.[9]

A second factor affecting stratification is the practice of lending and borrowing land between households. Families try to hold land in different areas to minimize risk in this unpredictable environment. All families depend on being able to borrow land from friends and relatives to meet seasonal needs, which will vary depending on factors of climate, labor availability, need for cash, and so on. In times of high flooding or heavy rainfall, farmers with extensive dhasheeg land but small upland holdings borrow highland plots; in times of drought, families with small amounts of dhasheeg land borrow dhasheeg plots. Requests to borrow land are seldom refused (I never heard of a request being denied), as the lender one season may need to borrow the next.[10]

It is important to note that the farmers of the middle Jubba are subsistence farmers. Here, stratification does not have implications for differential wealth accumulation; it does have implications for ability to ward off starvation. The main crop, maize, is grown virtually entirely for subsistence, while small amounts of sesame seeds are sold to itinerant traders for cash to meet household needs. Household income for all Loc households is very low, and there is no accumulation of wealth. The reciprocal obligations involved in lending and borrowing are to avoid starvation, not to amass wealth. Thus, no farmer would refuse a request to borrow; to do so might very well jeopardize the subsistence of the would-be borrower's family.

A third factor serving to equalize access to dhasheeg land is the existence of a swamplands "commons." The swampland (called *siimow*) is fed by underground water flow, which seeps into the area when the river is high. Agriculture in the siimow is very unpredictable, and thus the area was abandoned for permanent cultivation about thirty years ago. Since then, the adage that no one can claim private rights to land in the siimow has been in effect. Whenever the siimow is cultivable, farmers rush in to carve out and plant a parcel for the season. Arguments over boundaries are mediated by the village council, which oversees the seasonal use of the siimow. Planting is done on a first come, first served basis, and is equally open to all. The existence of an area of occasionally available dhasheeg-quality commonly held land provides another means by which a farmer with little dhasheeg land of his own can obtain seasonal access.

The existence of the siimow commons illustrates an important component of dhasheeg agriculture in the middle Jubba. Different tenure systems apply to different areas under different climatic conditions, and these systems change with seasonally varying land use needs. As discussed, individual tenure applies to the cultivation of dhasheeg parcels when the basins are dry. When the basins are full, however, the dhasheegs become commons; anyone can fish, harvest lily tubers (an important food) and utilize the water in the dhasheegs wherever they wish.[11] When full, dhasheegs are open to pastoralists for watering their livestock as well. When dhasheegs are dry, some aspects of communal control still operate. The village council oversees the perimeter of the dhasheegs, retaining the authority, on behalf of the community, to allocate land in these areas. Cooperative work groups are the primary source of labor for constructing drainage canals and earthen dams at the mouth of the dhasheeg. When machinery must be hired to block the entrance to the dhasheeg to prevent rising waters from entering and destroying maturing crops, everyone holding land in the dhasheeg must contribute toward the fee. And finally, the village council, and not some higher, regional authority, mediates disputes between farmers regarding their dhasheeg land. Land disputes between village farmers are never taken outside the village.

Government Intervention and Change

The management of local water resources in Loc is relatively informal and based on the kind of face-to-face interaction possible in a small community. There is no water-users association; rather, door-to-door collections are taken by a village council representative if money is needed to hire a tractor to create a drainage ditch or retaining wall. The only formal aspects of the system are the village council's authority to distribute land in dhasheegs and to mediate boundary and ownership disputes, particularly in the siimow. Because it is small-scale, localized, self-contained, and dependent on natural river flow with only limited human intervention, dhasheeg agriculture is a relatively simple, household-based form of irrigated agriculture. Under local management, community interests and individual interests in dhasheeg agriculture are seldom in opposition, as two primary characteristics of dhasheeg agriculture are equity and risk reduction.

In the late 1980s, state and foreign intervention in the indigenous agricultural practices of middle Jubba Valley farmers began to transform important aspects of local water use and management. Until then, the subsistence orientation, the practice of borrowing and lending land, and the maintenance of a swamplands commons ensured that all community members had access to vitally important dhasheeg land. The system minimized risk and maximized flexibility in meeting seasonal land needs, and it ensured access to flooded land for a variety of uses. The land registration law (Law No. 73), passed in 1975, was just beginning to affect farmers in the middle Jubba ten years later. The law allowed only one parcel of land per family, disallowed the trading, lending, and selling of land, removed the authority of the village council in land allocation and dispute mediation, and did not recognize the right of a community to hold areas of farmland in common. If enforced locally, the land law would have made many of the historically developed strategies of land use and land tenure in middle Jubba communities illegal.

The most significant effect of the land law was to highlight the importance of irrigable land in an era of significant foreign agricultural development projects. Widely known plans for irrigation development in the Jubba Valley encouraged land speculation by individuals not from the area, introducing new forms of stratification between politically savvy outside investors and illiterate local farmers dependent on illegal (according to the land law) land use strategies. By 1988, for example, outside speculators had claimed significant amounts of riverbank land, and one man had registered the bulk of the siimow in Loc, thus obtaining the legal right—unexercised in 1988—to bar local farmers from continuing to cultivate there.

With the fall of Somalia's government in 1991, a new avenue of land claims has emerged. Resistance fighters and militiamen are claiming irrigable land through armed force, superseding the claims of local farmers as well as speculators who had received title in the 1980s (African Rights 1993). Resolving the resulting hierarchy of overlapping claims to valuable irrigable land will undoubtedly emerge as one of the most significant issues underlying Somalia's ability to re-establish its productive agricultural base. If international monitoring organizations or a future Somali state have any role to play in this resolution, it will be to ensure local community involvement in the determination of effective and equitable dhasheeg land use and land tenure patterns.

Acknowledgments

Funding for the fieldwork upon which this paper is based was provided by the Land Tenure Center, University of Wisconsin, and the Jacob Javits Foundation. This paper was first conceived while I was a resident scholar at the School of American Research, and I thank them and the other institutions for their support. I express my gratitude to Tad Park, who was instrumental in helping me develop some of the ideas discussed here.

Notes

1. A very high flood in May 1987 covered most of the doonk land of the mid-valley village of Loc, where fieldwork was conducted. Yields on doonk land planted after the

floodwaters had receded were high, as this is the time when doonk produces best. That season farmers reported sesame yields of 0.72 quintals per *darab* (1 hectare equals 6 darabs) on doonk farms and 1.1 quintals per darab on dhasheeg farms. Maize yields were running about equal, although ten parcels of dhasheeg land with promising crops had not been harvested by the conclusion of my field research. In all probability, when all the harvests were in, dhasheeg farms most likely outproduced doonk farms for maize as well that year. It is important to remember that even if doonk land produces as well as dhasheeg land in seasons following a heavy flood, dhasheeg land consistently produces more than doonk land in years characterized by no flooding or little rain, which occur about four years out of five.

2. Although archaeological evidence indicates that humans have used the upper valley since the Middle Paleolithic Period (ca. 125,000 years B.P.), there is a gap in our knowledge of resource use in the valley from about 20,000 years ago to about 700 years ago. Turton's (1975) and Lewis' (1966) reconstruction of migrations, wars, and conquests of Somali and Boran (Galla) pastoralists over the past 700 years in this area, summarized here, is the most widely accepted thesis.

3. Arguments have been made, primarily based on the *Book of the Zanj*, that the great kingdom of Shungwaya had its capital on the lower Jubba and that Bantu-speaking farmers lived in the valley prior to the eighteenth century. The most important proponent of this view is Cerulli (1934). For a recent review of the argument that the region called Shungwaya, including the Jubba Valley, has "always . . . been inhabited," see Pouwels 1987. Most scholars now doubt the existence of the Shungwaya kingdom (see Turton 1975). Although it is possible that agriculturalists lived in the river valley centuries ago, European explorers traveling up the river in the mid 1800s took note of the dense jungle characterizing the valley, indicating that agriculture had not been practiced there for at least a century.

4. See Cassanelli (1982) for an excellent analysis of the emergence and growth of plantation agriculture on the Shabelle.

5. Menkhaus (1989) and Cassanelli (1989) are the best contemporary secondary works utilizing these documents. I rely on these sources for the present summary of early settlement.

6. Forest products such as wild animals, fruits, greens, honey, and fish continue to provide an important supplementary food source, especially during times of scarcity.

7. East African ethnic identity was initially very important in determining village composition in the lower Gosha (Menkhaus 1989). Newcomers would seek out and settle among people from their East African homelands. By the early years of the twentieth century, intermarriage, linguistic transition to Somali, religious conversion to Islam and other factors (discussed in detail in Besteman 1991) caused ethnic boundaries to blur and become less significant in Gosha identity.

8. Women rarely inherit land, and virtually never acquire it by other means. These figures thus more accurately represent the amount of land held by the male head of household, land that is used to support the household.

9. Under conditions of finite dhasheeg land, rising population, and the commercialization of agriculture it is, of course, likely that stratification of access could increase, after a certain threshold, due to competition for the valuable resource. By that time, inheritance may not be the key factor.

10. The period during which data was collected (1987–1988) was marked by a very high flood in May 1987. Fifty-three percent of Loc households reported lending upland plots during that season, and 40 percent reported borrowing upland plots. Following the flood, 28 percent of village households reported borrowing dhasheeg, and 25 percent reported lending dhasheeg. The lack of a one-to-one correspondence is due to the fact that some households borrowed or lent more than one plot.

11. Vondal (1987) reports a similar pattern in Borneo swamplands.

References Cited

African Rights

1993   *Land Tenure, the Creation of Famine, and the Prospect for Peace in Somalia.* Discussion Paper No. 1. October. London: African Rights.

Barker, Jonathon

1984   Politics in Production. In *The Politics of Agriculture in Tropical Africa,* ed. J. Barker, 11–31. Beverly Hills: Sage.

Berry, Sara

1984   The Food Crisis and Agrarian Change in Africa: A Review Essay. *African Studies Review* 27(2):59–112.

1989   Access, Control and Use of Resources in African Agriculture: An Introduction. *Africa* 59(1):1–5.

Besteman, Catherine

1991   *Land Tenure, Social Power, and the Legacy of Slavery in Southern Somalia.* Ph.D. dissertation, University of Arizona, Tucson. Ann Arbor: University Microfilms International.

Bloch, Peter

1986   *Land Tenure Issues in River Basin Development in Sub-Saharan Africa.* Land Tenure Center Research Paper No. 90. Madison: University of Wisconsin.

Brandt, Steven, and Thomas H. Gresham

1988   *JESS Report on Cultural Heritage Survey of the Proposed Baardheere Reservoir.* Report No. 26. Burlington, Vt.: Associates in Rural Development.

Brunken, Heiko, and Wolfgang Haupt

1986   The Importance of the Juba Valley for the Development of Agriculture in Somalia. In *Somalia: Agriculture in the Winds of Change,* eds. P. Conze and T. Labahn, 165–87. Saarbrucken: Epi Verlag.

Cassanelli, Lee

1982   *The Shaping of Somali Society: Reconstructing the History of a Pastoral People, 1600–1900.* Philadelphia: University of Pennsylvania Press.

1989   Social Construction on the Somali Frontier: Bantu Former Slave Communities in the Nineteenth Century. In *The African Frontier: The Reproduction of Traditional African Societies,* ed. I. Kopytoff. Bloomington: Indiana University Press.

Cerulli, Enrico

1934   Gruppi etnici negri nella Somalia. *Archivio per l'antropologia e la etnologia* 69: 177–84.

1957, 1959, 1964   *Somalia: Scritti vari editi ed inediti.* 3 vols. Rome: Istituto Poligrafico
    dello Stato.

Cohen, John
    1980   Land Tenure and Rural Development in Africa. In *Agricultural Development in
        Africa: Issues of Public Policy,* ed. R. Bates and M. Lofchie, 349–400. New York:
        Praeger.

Commins, Stephen, K. Michael Lofchie, and Rhys Payne (eds.)
    1986   *Africa's Agrarian Crisis: The Roots of Famine.* Boulder: Lynne Reinner Pub-
        lishers.

Conforti, E.
    1954   Studi per opere de risanamentoe valorizzazione dei desick del Giuba, folio 1560.
        Florence: Centrodi Documentazione, Istituto Agronomico.

Downs, Richard, and Steven Reyna (eds.)
    1988   *Land and Society in Contemporary Africa.* Hanover: University Press of New
        England.

Dundas, F. G.
    1893   Expedition up the Jub River through Somali-land, East Africa. *The Geographical
        Journal* 1:209–23.

Food Early Warning System
    n.d.   (FEWS) records. Mogadishu, Somalia: Ministry of Agriculture.

GTZ (German Agency for Technical Cooperation)
    1984   *Deshek and Small- and Medium-scale Irrigated Agriculture in the Juba Valley.*
        Prepared by Agrarund Hydrotechnic GmbH, Esseu, Germany.

Haugerud, Angelique
    1989   Land Tenure and Agrarian Change in Kenya. *Africa* 59(1):61–90.

Lewis, Herbert S.
    1966   The Origins of the Galla and Somali. *Journal of African History* 7(1):27–46.

Linares, Olga
    1981   From Tidal Swamp to Inland Valley: On the Social Organization of Wet Rice
        Cultivation Among the Diola of Senegal. *Africa* 51:557–95.

Menkhaus, Kenneth
    1989   Rural Transformation and the Roots of Underdevelopment in Somalia's Lower
        Jubba Valley. Ph.D. dissertation, University of South Carolina, Columbia.

Netting, Robert McC.
    1993   *Smallholders, Householders: Farm Families and the Ecology of Intensive, Sustain-
        able Agriculture.* Stanford: Stanford University Press.

Okoth-Ogendo, H. W. O.
    1976   African Land Tenure Reform. In *Agricultural Development in Kenya: An Eco-
        nomic Assessment,* ed. J. Heyer, J. K. Maitha, and W. M. Senga. Nairobi: Oxford
        University Press.

Park, Thomas
    1988   Indigenous Responses to Economic Development in Mauritania. *Urban Anthro-
        pology* 17(1):53–74.

1992   Early Trends Toward Class Stratification: Chaos, Common Property, and Flood Recession Agriculture. *American Anthropologist* 94(1):90–117.

Pouwels, Randall

1987   *Horn and Crescent: Cultural Change and Traditional Islam on the East African Coast 800–1900.* Cambridge: Cambridge University Press.

Reyna, Steven

1987   The Emergence of Land Concentration in the West African Savanna. *American Ethnologist* 14(3):523–41.

Richards, Paul

1983   Ecological Change and the Politics of African Land Use. *African Studies Review* 26(2):1–72.

Scudder, Thayer

1975   *Ecology of the Gwembe Tonga.* Manchester: Manchester Press (1960).

1992   *African River Basin Development.* Boulder, Colo.: Westview Press.

Tillman, Robert

1987   Environmental Issues: Baardheere Dam. Jubba Environmental and Socioeconomic Studies unpublished paper, available at Associates in Rural Development, Burlington, Vt.

Turton, E. R.

1975   Bantu, Galla, and Somali Migrations in the Horn of Africa: A Reassessment of the Juba-Tana Area. *Journal of African History* 16(4):519–37.

Vondal, Patricia

1987   The Common Swamplands of Southeastern Borneo: Multiple Use, Management, and Conflict. In *The Question of the Commons: The Culture and Ecology of Communal Resources,* ed. Bonnie McCay and James Acheson, 231–49. Tucson: University of Arizona Press.

Watts, Michael

1983   "Good Try Mr. Paul": Populism and the Politics of African Land Use. *African Studies Review* 26(2):73–83.

Watts, Michael, and Robert Shenton

1984   State and Agrarian Transformation in Nigeria. In *The Politics of Agriculture in Tropical Africa,* ed. Jonathon Barker, 173–204. Beverly Hills: Sage.

Zoli, Corrado

1927   *Oltre guiba.* Roma: Sindacato Italiano Artgrafiche.

4

The Dry and the Drier
Cooperation and Conflict in Moroccan Irrigation

John R. Welch

Successful reconstructions of human social evolution and plans for the future of an overpopulated globe depend, at least in part, on understanding stress and the patterns of conflict or cooperation it provokes. In this chapter I examine a stress familiar to agrarian populations in arid regions—water scarcity—as it is dealt with in four irrigation systems in the geographically unified, but ecologically and culturally diverse region surrounding the central Moroccan town of Sefrou (figure 4.1). The region's ethnographic records document irrigation strategies ranging from traditional, communally owned systems in the relatively wet, but infertile mountains, to larger, market-oriented systems fraught with competition over water. Responses to water deficits in the four systems appear to vary with the frequency and intensity of the stress, and cooperation is as common a response as conflict. These findings support a critique of Geertz's (1972) classic comparison of irrigation in Bali and Morocco, "The Wet and the Dry," wherein the author asserts that "agonistic individualism" is a Moroccan "cultural presupposition," an invariant aspect of the nation's human interaction that derives from water scarcity (this claim is expanded in subsequent work by Geertz, Geertz, and Rosen 1979 and Rosen 1984).

Beyond criticizing Geertz, I examine the structure of the Sefrou drought response. In the descent from the smaller, wetter upland systems to the larger, drier lowland fields, increasing water scarcity causes greater competition until, in the lowest and driest system, cooperation re-emerges as a dominant theme in stress mitigation and water use. Beyond the descriptive level, these patterns of unselfish cooperation and entrepreneurial competition appear to be structured hierarchically. Increasingly severe stresses provoke riskier and more costly responses; higher order attempts at stress mitigation tend to limit the effectiveness of lower-order responses. Each step up the hierarchy of response options carries exposure to new risk along with the promise of stress reduction. The concept of hierarchical response to living system perturbations developed in Slobodkin's (e.g., 1968) work provokes consideration of the temporal stability of

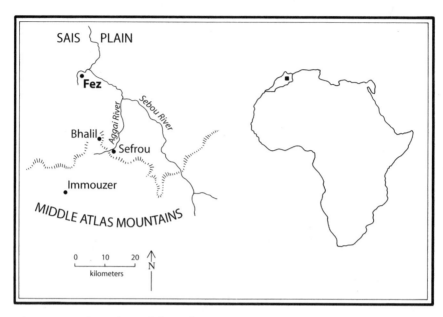

Figure 4.1   Location of the Sefrou region, Morocco.

agrarian adaptations to arid regions. The severe decline in Moroccan food production that accompanied the recent drought (1981–1984) increases the study's "real world" relevance. Ethnographic and archaeological research into the range and effectiveness of drought-coping mechanisms in arid lands' irrigation systems is needed.

The ecological and evolutionary framework adopted in this study differs from valuable work by Pascon (1977), Bouderbala et al. (1984), Hammoudi (1985), and Mahdi (1986). These analyses of the allocation of critical resources in irrigation systems point to economic and political histories as the primary determinants of behavior patterns. Data for a comparable analysis of the Sefrou region have not been obtained.

Sefrou Agrarian Ecology

The Sefrou region's environment is "characterized by a highly irregular and uncertain set of climatic conditions and a wide variety of microenvironments" (Geertz et al. 1979:8; see table 4.1). There are about 260 frost-free days at 1500 meters above sea level. Annual precipitation ranges from 300 to 1100 millimeters and occurs mostly from November through April;

Table 4.1 Environmental characteristics of the Sefrou region, Morocco.

	Mountains	Piedmont	Lowland plains
Elevation	1100–2500 meters	500–1100 meters	100–500 meters
Hydrology	Abundant surface water in springs, streams, lakes	Numerous, irregularly spaced springs	Surface water confined to numerous wadis, a few perennial streams
Settlement	Hamlets and camps near water, pasture	Villages and camps around oases and shrines	Towns and villages along water courses
Land Use/Users	Transhumant Berber pastoralists supplement economy with dry and irrigated cereal farming	Settled Berber pastoralists and irrigated agriculturalists with limited dry farming; some Arab population	Irrigated horticulture, some dry farming; urban markets; ethnic diversity
Limitations	Less than 1/5 of land arable; stony soils	Variable stream and spring flow; soils stony in places	Variable stream flows; socio-political factors
System	Ait Sidi Musa	Ain Sultan	Sefrou and Lower Aggai

midsummer is sometimes totally dry. The marked spatial, seasonal, and interannual variation in rainfall that Geertz identifies as the source of the confrontational nature of Moroccan socioeconomic intercourse is apparent (see also Kacemi 1992).

Social Patterns

The town of Sefrou occupies an escarpment that separates the lower piedmont from the wetter, upland plains and the zone of Berber domination from that of government control (Vinogradov 1974:38). Berber farmer-pastoralist economies dominate the piedmont and mountains. Some Berbers, along with Arabs, Jews, and Europeans, mingle in the more populous, fertile, economically specialized, and arid lowlands. Most of this discussion focuses on Berber ecology, but Arabs and others also participate in lowland irrigation. The two Berber subtribes examined here, the Ait Yusi (population 60,000 in 130 villages) and the Ait Mguild (population 74,000), are territorial groups with segmentary, patrilineal-patrilocal organization (see Chiapuris 1979; also Vinogradov 1974; Miller 1984).

Chiapuris (1977, 1979:58, 63) sees a Middle Atlas Berber unity derived from linguistic, cultural, and environmental commonalities and a previous concern with territorial defense.[1] Changing boundaries between groups serve to adjust populations to local resources, especially farmland and pasture (Chiapuris 1979:37). Kinship remains important in determining group affiliation, but membership today is more open. Territorial borders, once closely monitored and hotly contested, now change constantly with governmental decrees and private land transactions (Rosen 1979:24).

Agricultural Patterns

The people of Sefrou make a living through a combination of goat and sheep pastoralism, market and subsistence farming, and some manufacturing and trading. Agriculture has expanded in the last century as pastoral transhumance has been constrained by governmental policies and environmental degradation (Chiapuris 1979). The Ait Mguild cultivated 7,300 hectares in 1936; by 1950, this figure had grown to 33,500 hectares (Raynal 1960:305).

Four classes of agricultural land are found in the area: irrigated fields, dry plowlands, pastures, and less productive "no man's lands" separating Berber territories (Chiapuris 1979:68). Wheat is the staple crop, but a wide range of grains and vegetables is cultivated. Lowland plots are typically worth more than highland fields of the same size because of their easier access to markets, deeper soils, and longer growing seasons. Similarly, because irrigated fields can be planted more frequently and produce good harvests more reliably than dry upland farms, the former are typically worth at least twice as much. Shorter fallow cycles are possible on irrigated land due to nutrients brought by irrigation, fertilizers introduced using movable corrals, and to more common intercropping. The impossibility

Table 4.2 Estimated seed inputs and yields (kg/ha) in the Sefrou region, Morocco.

Crop	Seed input	Lowlands				Highlands			
		Irrigated fields		Dry fields		Irrigated fields		Dry fields	
		Yield	Ratio	Yield	Ratio	Yield	Ratio	Yield	Ratio
Hard wheat	90	540	1:6	360	1:4	630	1:7	450	1:5
Soft wheat	90	540	1:6	540	1:6	630	1:7	450	1:5
Barley	92	525	1:5	735	1:7	525	1:5	420	1:4
Maize	44	315	1:7			270	1:6		
Average		400	1:6	545	1:5.7	514	1:6.25	440	1:4.7

Source: Data from Boyle 1977:121–127; see also Chiapuris 1979:75.

of dry field double cropping means that—holding field size constant— wet fields commonly produce at least twice as much per year.

Boyle's (1977) study of the Ait Mguild illustrates patterns in Berber land tenure and transhumant cultivation. Table 4.2 describes the two cultivated territories of the Aīt Sīdī Musa clan. Only six of the thirty households that Boyle studied had irrigated land at the clan's lowland farms; twenty-seven had irrigated fields in the highlands. Poorer households had less total land and, proportionally, even less irrigated land. Furthermore, smaller landholders used shorter fallow cycles.

Even discounting irrigation system construction costs, labor input for irrigated fields is far greater in comparison to labor required in dry land farming. Dry fields need few water management inputs, receive less fertilizer, and are tilled less. Wheat, fava bean, and lentil seed are broadcast on dry fields prior to a second plowing. Dry barley fields are plowed only after broadcasting. Irrigated fields destined for maize, chickpeas, or wheat are preplowed. On lowland farms, where vegetation invades fallow lands more quickly than in the highlands, additional labor is required for plowing and weeding. The last notable difference in labor inputs is that farmers must descend periodically from summer herding camps to weed, water, and harvest lowland irrigated fields.

For a given harvest, the yields from wet and dry fields may be comparable. The critical difference is that the irrigated farms can be planted soon after harvest and be expected to produce equally consistently without fallow. Yields from dry fields drop precipitously when the soil is not rested. Interestingly, Berber farmers' reliance on irrigated lowland farms seems to have occurred in the absence of absolute land shortages. Instead, this form of labor intensification may buffer circumscribed pastoralists against social and climatic uncertainty.

Irrigation Patterns

Two distinct types of water distribution are used in the Sefrou region (see Geertz 1972:34). In the *mubīh* or "queue" distribution system, water and land are inseparable (see Glick's Syrian type; 1970:230). A landholder's water right is derived from kinship, and cannot be transferred independently of land. Nor can the right holder's place in the queue be used to irrigate another plot. This inflexibility permits no commerce in land and water as separate commodities, thus encouraging continuous, intensive plot cultivation.

The second approach, *l-māa' bi-sāà* (water by the hour) is comparable to Glick's (1970) Yemenite type of irrigation and appears more attuned to market forces. Rights to the water source for a specified period are held separately from land and may be transferred (through purchase or loan) for use on any plot. The volume delivered is a function of water flow over the time owned, and farmers engage in entrepreneurial maneuvering to obtain sufficient water, especially during droughts. Field condition (i.e., degree of desiccation) is only considered in l-māa' bi-sāà distribution when water is abundant. In contrast, in queue systems, the length of time a landholder has access to water is dependent on the flow volume and the size and condition of the plot. Right holders at the head of the queue are, especially in dry years, in an enviable position to monopolize the irrigation flow.

Four Sefrou Irrigation Systems

With this information as background, each system's organizational and technical means for capturing and allocating water and for coping with shortages can now be examined. Table 4.3 summarizes the four systems, arranging them from dry to driest, from highest in elevation to lowest, and from relative socioeconomic homogeneity to heterogeneity.[2]

Table 4.3 Characteristics of four irrigation systems of the Sefrou Region, Morocco.

	Ait Sidi Musa	Ait Sidi Lahcen	Sefrou Oasis	Lower Aggai
Precipitation	dry (1100 mm)	—	—	driest (200 mm)
Elevation	high (1300 m)	—	—	lowest (500 m)
System size	smallest (147 ha)	large	largest (625 ha)	large
Source	stream	springs	Aggai River	Aggai River
Distribution	queue	timed	combination	combination
Market role	minimal	olives only	primarily cash	vegetables
Cash crop	crops (vegetables)	olive trees		
Population	low	medium	highest	high
Settlement	camps	village around springs and shrine	Sefrou (30,000)	villages outside of Sefrou
Land and water	inalienable: kinship-based	alienable: based on fictive kinship	alienable: bought and sold	alienable: bought, sold, and shared
Authority	1 part-time jārí	1 full-time jārí	5 full-time jārí	2 full-time jārí
Social diversity	lowest	—	—	highest
Disputes	rare	occasional	constant	common

Upland Irrigation Systems of the Ait Sidi Musa Clan

About half of the Sidi Musa clan's highland farms are irrigated with simple ditches leading from permanent streams. Irrigation is preferred, but most families also dry farm. Water distribution involves breaching one of the shallow ditches to allow water to run down cultivated slopes. Boyle (1977: 122) estimates that 30 to 40 percent of the flow is wasted through inefficient distribution, but notes that there is enough water to meet most needs in most years.

Informal associations of farmers using particular ditches administer the Ait Sidi Musa system. The size of the associations ranges from four to twelve farmers, depending on ditch length. The group selects an irrigation chief (*jārí*) from the membership. The jārí should reside near the ditch and be respected, energetic, and judicious. In addition to his own farming, the jārí delegates maintenance tasks, supervises water distribution, and mediates disputes. If jārí mediation is unsuccessful, lineage elders, the tribal council, and—if the conflict remains unresolved—Morocco's civil authorities and regional and appellate courts may be called upon in turn. This is the first example of hierarchical response to stress. The means for dispute resolution can be seen as a series of increasingly costly and risky responses to persisting problems.

Sultan's Spring (Ain Sultan) Irrigation System

The second system is lower and slightly drier. The Ain Sultan spring is the water source for the approximately 400 households that make up the village of Sidi Lahcen. Each household relies on an average of 2 hectares of dry and 1 hectare of irrigated land. Rosen (1979:31) observes that "most of the villagers attempt to keep three or four separate plots. . . . This distributes the risk of crop failure over several microecological niches." The size and quality of the plots are, in general, evenly distributed among the households.

Irrigation is the foundation for the Sidi Lahcen economy. Most subsistence crops are dry-farmed, but the 20,000 olive trees—the primary source for cash—depend on irrigation: "Practically all the trees and land at Sidi Lahcen are private property. . . the olive crop is essential for the purchase of goods from outside the valley" (Rosen 1979:31). The Sultan's spring is led to fields by a 6-kilometer ditch. The l-māa' bi-sāà distribution cycles through ten sixteen-hour days and twelve eight-hour nights. Turns in the cycle may be as short as fifteen minutes, and water must be available for the next user when the turn ends. That the allocations are timed in a fixed cycle suggests the evolution of "water by the hour" distribution from a queue type of system.

Negotiations to minimize water movement between plots and to assure prompt delivery are continuous. The haggling employs a vocabulary based on farmers' putative genealogical links to the village founders—the developers of the irrigation system. Four of the ten named days are reserved for alleged founder descendants in upper villages, two days for descendants living in lower villages, and four for lineages lacking ties to founders. Each of these days is further partitioned by clan and family membership. Rosen (1979:33) notes that this "conceptual system projects an overarching level of relatedness . . . that serves to order the basic structure of the irrigation system at the same time it allows maximal adjustment to ecological, political, and personal necessity."

When there is consensus to begin the summer's irrigation, the water holders meet, elect a jārí, and establish a starting point for the delivery sequence by lottery. Each water right holder is responsible for maintaining a stretch of the primary ditch proportional to his water rights. Land holders lacking irrigation rights have no maintenance duties. Because unirrigated land produces little in most years, water assets are, figuratively as well as literally, perceived as more liquid—and thus more valuable. The dispute resolution hierarchy is similar to that described for the Sidi Musa system.

Town of Sefrou Irrigation System

Lower and drier still is the Sefrou Oasis system. The five primary canals leading from the Aggai River in 1955 irrigated 1260 plots (Rosen 1979:115, 118). Each canal feeds a discrete set of fields and is administered separately. Both land and water are privately, rather than communally, held. The five irrigation chiefs are appointed by the Sefrou pasha and are occupied full-time from March through October managing ditch maintenance, the blend of timed and queue distribution, and frequent disputes over flow allocation. Unlike jārí for the two upland systems, these men have legal authority to settle disputes, delegate maintenance tasks, and impose penalties for deviations from established procedures. For these reasons, and because they are responsible to the pasha, these jārí are accorded greater deference than in the uplands, and often use this status to their advantage in other social and economic domains.

Individual plots receive what their owners (and whoever feels like contributing an opinion) consider a thorough watering in a mubīh cycle, but the river is allocated to each primary canal by the hour, using a l-māa' bi-sāà variant. Water shortages prompt greater interest in allocation by the individual farmers and subject jārí decisions to increasingly thorough scrutiny. Perhaps because of social heterogeneity and market penetration, the Sefrou irrigation system contains few kinship-based sources of organizational variability. This feature of the Sefrou system accounts for the observation that, despite a reliable water supply, "disputes are common and quite heated" (Rosen 1979:61): "An overall moral and legal code serves as the context in which individuals contract and contend with one another, adjusting their personal needs more through direct confrontation—a kind of agonistic individualism—than through elaborately constructed group relations."

Unresolved disputes may be brought up at plot owners' meetings. If the dispute persists or escalates, an honored elder may mediate. Continuing dissension may then be taken, in turn, before the Sefrou pasha, the local civil governor, and the regional and national courts. Each step up this hierarchy brings greater public attention and increased potential losses of local autonomy and individual status.

The Lower Aggai Irrigation System

Below Sefrou the dwindling and increasingly saline Aggai River is virtually consumed by three more canals. Because of flow unreliability, all water is directed to the largest of the three canals (Al-glet) one year and divided between the two smaller canals (Beni Mansour, Dar Attar) in alternating years. Cash-economy forces are apparent: Olive trees must be irrigated at least enough to sustain them through years when their owners lack irrigation rights. Plots on the canal with rights in a given year are typically planted in maize and vegetables; hardy cereals are planted in plots slated to be dry. Farmers with irrigation rights during drought years can realize large profits on scarce, and, therefore, lucrative garden crops. These factors encourage efficient use of the meager flow by farmers with rights for the year.

Mubīh distribution is the rule, but important exceptions occur. During wet years, when the two smaller canals share the flow, Dar Attar receives two-thirds of each day's water, Beni Mansour the remaining third. Similar partitioning occurs on the Al-glet canal, with secondary ditches receiving proportionally smaller allocations. Compared to the other three systems, the more prevalent water scarcity in the Lower Aggai system means less reliable harvests. Farmers cope with this by constantly adjusting the timing of irrigation, relaxing the alternating legal prerogative to the river, regularly sharing scarce water, and stooping to elaborate forms of deceit and theft (Rosen 1979:121–122).

Cooperation, Conflict, or Combination?

Geertz's comparison of Balinese and Moroccan irrigation ecology concludes that, "As a chameleon tunes himself to his setting . . . a society tunes itself to its landscape. . . To connect the restless irregularity of much of Moroccan life, the tense expectancy and the aggressive opportunism of it, to the uncertain, capricious climate is not, therefore to yield to vulgar materialism, for it is in part that climate that projects the aura of irregularity and tension" (Geertz 1972:38). Reminiscent of "national character" studies by Mead and Benedict, Geertz's poetic notion that "agonistic individualism" derives from Morocco's aridity glosses over the links among social and environmental variables. In so doing, Geertz (1972:37) disregards his own good advice (1972:37): "environment is but one variable among many. . . . And it is one whose actual force must be empirically determined, not a priori declaimed." With interpretation as both the point of departure and the destination for Geertz's study, there is no appeal to epistemological principles (see also Netting 1982; Foster 1986). Furthermore, the assertion that an entire culture is basally eristic could be construed as racist.

Cooperation in Sefrou Irrigation

"The necessity for an ordered, predictable system of water allocation has divisive as well as cohesive potentialities" (Millon, Hall, and Diaz 1962: 513). Water scarcity can and does foster both cooperation and conflict in irrigation systems (see Maass and Anderson 1978; Bolin 1990), and the four systems discussed here are exemplary in this regard. Geertz and Rosen, however, focus on the discordant effect of insufficient water and imply that conflict should increase as water grows scarce.

Long-term studies of individual systems are required to fully assess this idea, but the comparisons drawn here offer some insights. Water scarcity and conflict increase concomitantly through the first three systems. But competitive behavior peaks (in the Sefrou oasis) before the low point of water scarcity (Lower Aggai) is reached. It is in the most drought-prone system that cooperation resurfaces. Competition remains fierce in some Lower Aggai arenas, but it cannot escalate without the loss of valuable olive trees. Because the Lower Aggai olive trees require summer irrigation merely to survive, farmers from the division slated to be dry for a particular summer rely on the year's wet division to relax their rights to the entire flow. If one division asserted their prerogative, they could expect the same treatment during the next drought. The olive trees would go thirsty; some would die, and no one would be richer or wiser for having exercised their rights.

Also central to the Geertz and Rosen arguments is the assumption that water is treated as a commodity throughout Morocco. Geertz (1972:36) asserts, "private ownership of water is the organizing principle, a principle developed to levels of legal complexity which stand in marked contrast to the technical simplicity of the actual system." Because Geertz provides few historical or political data to support this argument (cf., Geertz 1963), the attribution of agonistic individualism to privatization is unconvincing. Another problem in this regard is that neither Geertz nor Rosen consider the implications of mubīh distribution, in which individual's irrigation rights are usufruct and water and land are inalienable. Mubīh is the rule in the highest system, where self-promoting maneuvers are discouraged through collective water ownership. The land is owned by individuals, but is virtually worthless without water. Irrigation rights are unavailable to farmers who fail to fulfill obligations to the collective. The Lower Aggai system indicates that queue-type allocation can serve diverse individual and collective goals even in a large system.

Cooperation rather than individualism prevails in all successful irrigation systems (Lees 1989; Bolin 1990), and a consideration of the many forms of cooperation central to Moroccan irrigation further challenges Geertz's lax methods and interpretive conclusions. In Sefrou, communal crop harvesting and transport, irrigation facility maintenance, and religious brotherhoods create multidimensional ties between individuals (for another Moroccan example, see Herzenni 1994). These ties supplement kinship links in systems that might be considered more traditional; they replace or coopt blood relationships in social contexts less structured by kinship.

Even as farmers retreat from kinship-based cooperation, economic relations (even those between members of unrelated clans) continue to be processed through a kinship idiom (see Boyle 1977). Lineage, religious, and tribal affiliations persist as significant determinants of patterns of economic interaction and land ownership in the Sefrou region, especially in circumscribed oases and pastures. Furthermore, economic gains invariably radiate along ties based on such affiliations. There are noncontractual, status-based aspects of all of these interpersonal ties, and these relationships help to define cooperative and competitive responses to water scarcity and other stresses. Competition is but one side of Sefrou irrigation, one perspective on a complex situation. Agonistic individualism may be only the most striking aspect (to an outsider) of the region's socio-economic process. Geertz rightly considers water scarcity as a powerful determinant of patterns in behavior and world view, but his failure to examine cooperation and other possible determinants invites criticism.

Ecology, Evolution, and Adaptive Response

Regardless of their focus on material conditions, symbol systems, or other factors influencing behavior, anthropologists' conclusions should be well grounded in empirical data. Geertz's interpretive approach *assumes* adaptation in its search for webs of meaning. In contrast, ecological and evolutionary approaches generally avoid submerging empirical observations in the murk of semantic coherence. Such positivist studies grapple with the complex present — often in pursuit of information useful in inferring aspects of past behavior or discerning future needs (see Raikes 1967). People do not live by the integration of their symbolic structures alone, and it seems reasonable to request that understanding and promoting adaptation be given priority over elegant interpretive excursions (see Netting 1982).

The concept of adaptation is common ground for ecological and evolutionary anthropology. Slobodkin's (1968:202) definition of adaptation — a living system is adapted to the extent that it maintains homeostasis in response to stresses (selective pressures) without sacrificing the capacity to respond to subsequent stresses — brings static ecosystems into the evolutionary realm of selective pressures and differential survival. Slobodkin's predictive theory of adaptation holds that organisms faced with growing (or undiminishing) stress will employ more costly responses. This framework provides both theory and method for analyzing stress response (see Minnis 1985).

The Sefrou data constitute an appropriate case for applying Slobodkin's framework for at least three reasons. First, water availability is a significant problem. Precipitation shortfalls result in poor harvests for that year and can seriously deplete the groundwater reserves upon which subsequent harvests depend (Kacemi 1992). Second, irrigation is central to Morocco's food supply and to the long-term success of individuals, households, and communities. Farmers surveyed by Boyle (1977:149) "claimed a twofold to threefold difference in yields on irrigable land between good and bad years, a fivefold to sixfold difference on dry land." Although groundwater-based irrigation systems (e.g., Ain Sultan) are preferred due to greater resistance to drought, most systems rely on the less predictable surface flows. Third, each of the four Sefrou systems includes hierarchical structures for response to water scarcity. Drought-resistant cultigens are planted. Smaller amounts of land are plowed and seeded. Water may be stored "in the soil" by avoiding tillage. Water capture and delivery efficiency are enhanced by cleaning canals and springs, building higher and sturdier banks to curb losses to evaporation and infiltration, or irrigating through the night. The length of turns can be more closely monitored or decreased across the board. Rosen (1984:63–65) even describes a hierarchy of prayers employed to deal with water scarcity. The sequence runs from everyday preemptive requests, through stronger cycles to interrupt dry spells, and into religious petitions so forceful that the people recognize the risk that destructive floods, rather than healing rains, may answer their appeals.

When these responses to water scarcity fail, elaborate forms of theft and deceit may emerge; increasingly formal and institutionalized means of conflict resolution may be called upon. Such a cycle of escalating responses to stress may eventually entail radical and irreversible social alteration. Each of the Sefrou systems includes means for mediating disputes over land and water, first by individuals, then by jārí, then by local notables and authorities, and finally by regional and national courts. These variations on the hierarchical response theme underscore the wide range of options available to farmers concerned with water scarcity. From this perspective, the Sefrou systems seem generally well adapted in terms of their stress-coping capacity.

Hierarchical Response and Differential Survival

Prolonged drought is clearly among the many challenges separating sustainable strategies from those less well-adapted. Documenting the variable effectiveness and resilience of subsistence strategies where water scarcity is a primary limiting factor contributes insight into the development of human economic strategies and our ability to feed growing populations in arid lands. Anthropologists need better techniques for understanding the evolutionary bottlenecks that Weins (1977) calls "crunch periods," and Slobodkin's (1968) concept is one promising avenue.

The Moroccan data indicate the utility of hierarchical response for organizing and comparing information on irrigation systems and suggest the potential benefits of expanding the analysis through ethnographic and archaeological studies. As analytic units, irrigation systems are superior to villages because of the clear connections to soil, climate, hydrology, technology, and social groups (Brunhes 1902; Kelly 1983). Ties to conflict and risk management structures are also apparent, particularly in arid lands (see Steward 1955; Lees 1974; Hunt and Hunt 1976).

A detailed study of irrigation system responses to water scarcity might profitably employ a Slobodkin framework for data collection and analysis. The results of long-term fieldwork could be integrated with information on variations in climatic, hydrological, and technological factors. Collecting detailed data on the magnitude, duration, and frequency of water scarcity, along with the costliness, effectiveness, and implications of responses would refine and quantify definitions for variables central to Slobodkin's theory.

Such an analysis of irrigation ecology would be valuable for several reasons. First, if Slobodkin's approach proves useful, anthropology would benefit from a definition of adaptation that integrates ecological and evolutionary approaches appropriate for analyses on a human scale. Second, because irrigation systems are material representations of social institutions and behavior patterns, archaeologists involved in their study could expand and strengthen inferences about ancient institutions and behaviors. Even acknowledging the importance of historical depth in analyses of irrigation systems, a vast gap remains between the pressing need to study the differential success of agrarian strategies and the scattered scraps of crockery, cutlery, and architecture that make up the archaeological record. We know little about the demographic, economic and organizational information encoded in irrigation system layout and technology. Bridging this gap will entail carefully focused fieldwork with living, functioning irrigation systems — irrigation ethnoarchaeology: "There is a great need for mapping present day systems . . . and relating to them the social systems concerned. . . . Having done that, it should be possible to determine whether a social organization receives its essential structure from factors outside irrigation needs" (Downing and Gibson 1974:x–xi).

Archaeology could, through this type of research, contribute to the study of sustainable adaptation, thus distancing the discipline from an earlier, more academic and antiquarian preoccupation with the splendors of ancient civilizations. Long-abandoned systems could provide cautionary tales as well as innovative strategies for dealing with nagging problems of fluctuating food and water supplies. In light of the severe drought that crippled North African agriculture in the early 1980s, research into extant irrigation systems should be tailored to complement development programmers' repertoire of technological and organizational strategies (Stockton 1988).

Concluding Thoughts

Irrigation agriculturalists have long attracted social scientists equipped with diverse methods, theories, and goals. This chapter has argued for the potential and need for research focused on the ecology and evolution of arid land agricultural systems. This focus, unlike the interpretive approach critiqued here, promises to help us to improve life where drought constitutes a prominent hazard. Too seldom are ecosystemic intricacies studied in conjunction with information on selective factors and adaptive responses in tightly defined behavior systems. This comparison has demonstrated the complexity—abstracted to the point of oversimplification by Geertz—of Sefrou irrigation. Slobodkin's theory of hierarchical response to stress has been useful in examining the Sefrou systems and merits further operationalization. The collection of quantified data on water scarcity and irrigation system coping mechanisms are appropriate directions for irrigation research in Morocco and elsewhere. The causes and consequences of variability in stresses and responses in living systems must be studied as a baseline for tracking the course of human socioeconomic evolution. Even if the data collected are inconsistent with Slobodkin's theory, the resulting technical and analytical methods will help to integrate ecological and evolutionary approaches.

Acknowledgments

This chapter is the result of research initiated in a 1986 cultural ecology seminar at the University of Arizona. The author is deeply indebted to the seminar leader, Bob Netting, and dedicates this work to the memory of this superb scholar and human being. Carol

Kramer, Tad Park, and Jonathan Mabry also provided useful comments on earlier versions of the chapter.

Notes

1. The antiquity of communal land tenure and its ties to "traditional" as opposed to market-oriented economies are mentioned several times in this chapter, but these issues are too complex to address adequately here (for historical background see Gellner and Micaud 1972).

2. The uneven quality of the data employed will not support quantitative comparison of the four systems' productivity and sensitivity to climatic variation.

References Cited

Bolin, Inge
    1990   Upsetting the Power Balance: Cooperation, Competition and Conflict along an Andean Irrigation System. *Human Organization* 49:140–48.

Bouderbala, N., J. Chiche, A. Herzenni, and Paul Pascon
    1984   *La Question Hydraulique*. Rabat: Institut Agronomique et Veterinaire Hassan II.

Boyle, Walden P.
    1977   *Contract and Kinship: The Economic Organization of the Beni Mguild Berbers of Morocco*. Ph.D. dissertation, University of California, Los Angeles. Ann Arbor: University Microfilms International.

Brunhes, Jean
    1902   *L'Irrigation: Ses Conditions Geographiques, ses Modes et son Organisation*. Paris.

Chiapuris, John Paul
    1977   *The Ait Ayash of the Central Atlas: a Study of Social Organization in Morocco*. Ph.D. dissertation, University of Michigan. Ann Arbor: University Microfilms International.
    1979   *The Ait Ayash of the High Moulauya Plain: Rural Social Organization in Morocco*. Anthropological Papers of the Museum of Anthropology No. 69. Ann Arbor: University of Michigan.

Downing, Theodore E., and McGuire Gibson
    1974   Preface. In *Irrigation's Impact on Society*, ed. T. E. Downing and McG. Gibson, ix–xi. Anthropological Papers No. 25. Tucson: University of Arizona Press.

Foster, S. W.
    1986   Review of *Bargaining for Reality*. *American Anthropologist* 88: 206–8.

Geertz, Clifford
    1963   *Agricultural Involution: The Processes of Ecological Change in Indonesia*. Berkeley: University of California Press.
    1972   The Wet and the Dry: Traditional Irrigation in Bali and Morocco. *Human Ecology* 1:23–39.

Geertz, Clifford, Hildred Geertz, and Lawrence Rosen
    1979   *Meaning and Order in Moroccan Society*. New York: Cambridge University Press.

Gellner, Ernest, and Charles Micaud (eds.)
   1972   *Arabs and Berbers: From Tribe to Nation in North Africa.* Lexington, Mass.: Lexington Press.
Glick, Thomas F.
   1970   *Irrigation and Society in Medieval Valencia.* Cambridge: Harvard University Press.
Hammoudi, Abdellah
   1985   Subsistence and Relation: Water Rights and Water Distribution in the Dra'a Valley. In *Property, Social Structure and Law in the Modern Middle East,* ed. A. Mayer, 27–57. Albany: State University of New York Press.
Herzenni, Ahmed
   1994   *The Cultural Economy of Technical Innovation in Semi-Arid Rural Morocco.* Ph.D., University of Kentucky, Lexington. Ann Arbor: University Microfilms International.
Hunt, Eva, and Robert C. Hunt
   1976   Canal Irrigation and Local Social Organization. *Current Anthropology* 17:389–411.
Kacemi, Mouloud
   1992   *Water Conservation, Crop Rotations, and Tillage Systems in Semiarid Morocco.* Ph.D. dissertation, Colorado State University, Fort Collins. Ann Arbor: University Microfilms International.
Kelly, William
   1983   Concepts in the Anthropological Study of Irrigation. *American Anthropologist* 85:880–86.
Lees, Susan H.
   1974   Hydraulic Development as a Process of Response. *Human Ecology* 2:159–75.
   1989   On Irrigation and the Conflict Myth. *Current Anthropology* 30:343–44.
Maass, Arthur, and Raymond L. Anderson (editors)
   1978   . . . *and the Desert Shall Rejoice: Conflict, Growth, and Justice in Arid Environments.* Cambridge, Mass.: MIT Press.
Mahdi, Mohamed
   1986   Private Rights and Collective Management of Water in a High Atlas Berber Tribe. In *Proceedings of the Conference on Common Property Resource Management, 1985,* 181–97. Washington, D.C.: National Academy Press.
Miller, James A.
   1984   *Imlil: A Moroccan Mountain Community in Change.* Boulder, Colo.: Westview Press.
Millon, René, Clara Hall, and May Diaz
   1962   Conflict in the Modern Teotihuacan Irrigation System. *Comparative Studies in Society and History* 4:394–521.
Minnis, Paul E.
   1985   *Social Adaptation to Food Stress.* Chicago: University of Chicago Press.
Netting, Robert McC.
   1982   The Ecological Perspective: Holism and Scholasticism in Anthropology. In

Crisis in Anthropology, eds. E. Adamson Hoebel, R. Currier, and S. Kaiser, 271–92. New York: Garland Press.

Pascon, Paul

1977   *Le Haouz de Marrakech,* 2 vols. Rabat: CURS; Paris: CNRS.

Raikes, Robert

1967   *Water, Weather, and Prehistory.* London: John Baker.

Raynal, Rene

1960   La Terre et l'Homme en Haute Moulouya. *Bulletin Economique et Social du Maroc* 24(86–87):281–346.

Rosen, Lawrence

1979   Social Identity and Points of Attachment: Approaches to Social Organization. In *Meaning and Order in Moroccan Society,* ed. C. Geertz, H. Geertz, and L. Rosen, 19–22. New York: Cambridge University Press.

1984   *Bargaining for Reality, the Construction of Social Relations in a Muslim Community.* Chicago: University of Chicago Press.

Slobodkin, Lawrence R.

1968   Toward a Predictive Theory of Evolution. In *Population Biology and Evolution,* ed. R. L. Lewontin, 187–205. Syracuse: Syracuse University Press.

Steward, Julian H.

1955   *Irrigation Civilization: A Comparative Study.* Social Science Monograph No. 1. Washington, D.C.: Pan American Union.

Stockton, Charles W. (ed.)

1988   *Drought, Water Management and Food Production.* Mohammedia: Kingdom of Morocco.

Vinogradov, Amal Rassam

1974   *The Ait Ndhir of Morocco: A Study of the Social Transformation of a Berber Tribe.* Anthropological Papers of the Museum of Anthropology No. 55. Ann Arbor: University of Michigan.

Weins, John A.

1977   On Competition in Variable Environments. *American Scientist* 65:590–97.

5

The Political Ecology of Irrigation in an Andean Peasant Community

Paul H. Gelles

Irrigation technology and water management, together with terracing, camelid herding, and the vertical control of different ecological niches facilitated the development of pre–Columbian indigenous empires in the rugged environment of the Central Andes. Today, irrigation remains a fundamental component of production in thousands of highland communities in Peru, Bolivia, and Ecuador. My view is that analysis of productive systems such as those found in highland irrigation requires that we go beyond technical, economic, or ecological considerations. These systems must be understood in terms of the cultural models that encode their implementation and the relationship of these models to political forces at the local, regional, national, and international levels. In this chapter I use such a perspective to understand the political ecology of irrigation in Cabanaconde, a large peasant community in southern Peru.

The way I use the concept of political ecology stresses that human interaction with the environment is mediated by political and cultural forms and forces.[1] Cabanaconde has made a sustained series of attempts to expand its hydraulic resources since at least the turn of the century; some attempts have been successful, others not. These communal initiatives, and local Andean systems of production in general, must be understood within the larger political economy and cultural politics of the Peruvian state. Huge development projects oriented to large-scale coastal agriculture contrast with the relatively small, often times non-existent, efforts to develop the hydraulic potential of the highlands. As I show in this chapter, the coastally based national economy's need for highland water has wide-ranging political effects at the local level.

The focus of my chapter is the local perception, management, and defense of water resources and other common property, and the ways that these local factors condition state and community relations. I examine water politics and community mobilization at three levels. First I address the cultural construction of water at the local level and the way this is tied to annual attempts by the community to expand its hydraulic resources.

Next, I explore regional factors that impede or facilitate the exploitation of hydraulic resources on communal territory. The final level of analysis concerns the manipulation of communal resources by the state and by an international consortium, and the local response to this.

I conclude with a discussion that places state and local initiatives to manage and expand hydraulic resources within a larger cultural politics between the coast and highlands in Peru. As my chapter demonstrates, local irrigation practices in the highlands, chronically ignored or challenged by the irrigation bureaucracies of the Peruvian state, are related to a wide range of semantic and social fields of community life. Maintenance of ethnic identity and resistance to the secularization, monetarization, and control of water management by the state are expressed in the continued use of local models of irrigation.[2] And as we will see in the case of the large, state-sponsored Majes canal, less subtle forms of peasant resistance and mobilization also challenge the prerogatives of the Peruvian state and "irrigation development."

Cabanaconde

Cabanaconde, located at 3270 meters above sea level in the Colca Valley (Province of Caylloma, Department of Arequipa), is the largest community of the Colca Valley (figure 5.1). The people of Cabanaconde, numbering upwards of 4000 people at present, are bilingual Quechua and Spanish speakers. The community is clearly the product of Peru's "colonial matrix" (Fuenzalida 1970), which brought together indigenous and European social, cultural, and political forms to constitute a new and unique entity. Today, the Cabaneños distinguish themselves ethnically from other groups in the southern Peruvian Andes, and even from other groups of the Colca Valley.[3] Although Cabanaconde, unlike many other highland communities in Southern Peru, has not been brutalized by the Peruvian military or by Shining Path, the reverberations of the dirty war make themselves felt there, as do many other political and economic forces.[4]

One hundred percent of Cabanaconde's agriculture is irrigated, the water coming from the Majes canal (described below) and the snow melt of Hualca-Hualca, a 6000-meter-high peak. Hualca-Hualca Mountain, alongside "Earthmother" and figures in the Catholic pantheon, is a principal deity and the object of much worship. Twenty-four hours a day for several months a year, irrigation water descends the Hualca-Hualca River

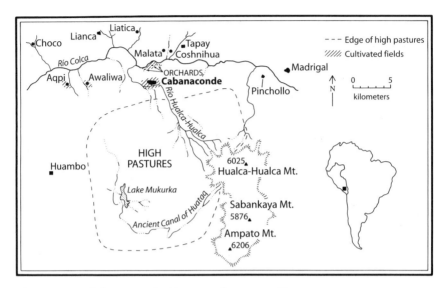

Figure 5.1 Cabanaconde, Peru, and surrounding area.

and passes directly through a series of canals to the more than 1200 hectares of terraced fields presently cultivated in Cabanaconde.

The perennial Hualca-Hualca River, together with the relatively gentle slope and warm microclimate of the Cabanaconde Valley, have always attracted extractive states. Archaeological studies reveal that the ethnic groups of this region had been part of at least two pan-Andean empires, Wari and Inka, before the arrival of the Spaniards (de la Vera Cruz 1987).[5] A large part of the pre–Inka agricultural infrastructure built in the area of Cabanaconde, especially canals and terraces, was expanded upon during Inka times (de la Vera Cruz 1987). Many of the canals, as well as hundreds of hectares of cultivated terraced fields, were abandoned after the Spanish invasion. During the early colonial period, the population was reduced by more than 80 percent.[6]

The population only began to recover rapidly in the mid-1800s. The number of Cabaneños more than doubled over the last century, rising from some 1,796 inhabitants in 1876, to 2,960 in 1940 and 3,421 in 1981 (Cook 1982:41, 84; Denevan 1987:17). The Ministry of Agriculture (1987) estimated that there were 800 families and a total population of 4000 in 1987.[7] Population pressure has led to outmigration. There has been a considerable amount of seasonal and permanent migration since at least the

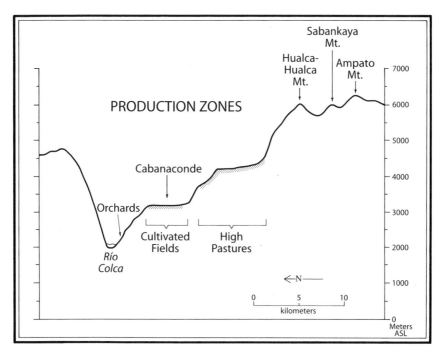

Figure 5.2 Production zones in Cabanaconde.

early part of this century, perhaps best evidenced by the Provincial Clubs of Cabanaconde operating in both Lima and Arequipa by the 1940s. Yet, Cabanaconde has experienced increased rates of outmigration since 1965, when a road reached the community. Today, there are large migrant colonies of Cabaneños in Arequipa, Lima, and Washington, D.C., estimated to have 1000, 3000, and 300 members, respectively. The provincial clubs formed by migrants in these cities are very much a part of the life of the community, and have intervened decisively in conflicts between the community and outside interests.[8]

Ecology and Irrigation

Cabanaconde lies on the arid west side of the Andes and is oriented toward the dry desert coast. The territory of Cabanaconde is environmentally diverse, with production zones ranging from 2,000 to 4,500 meters above sea level (figure 5.2). In the Colca Canyon at 2000 meters, the

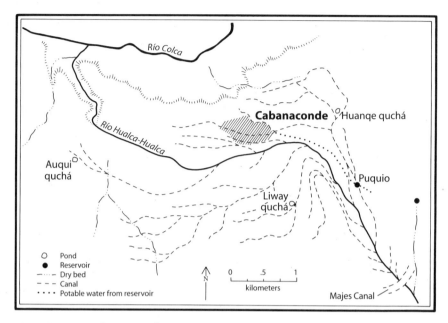

Figure 5.3   Cabanaconde irrigation system.

farmers cultivate fruit, alfalfa, and cochinilla (small insects that contain a red colorant that is sold commercially). In the high pastures, located at between 3800 and 4500 meters, alpaca, llama, sheep, and cattle herds are kept. These different ecological floors or "production zones" (Mayer 1985) are all within a day's walk of the community, constituting a classic case of "compressed verticality" (Brush 1977). The bulk of agricultural production takes place between approximately 2900 and 3350 meters, where the "Cabanita maize," famous for its taste and quality, is grown. There are over 1200 hectares of irrigated lands in the main fields, approximately three quarters of which are dedicated to this maize. All agriculture in Cabanaconde is irrigated, and irrigated land is the central source of wealth in the community, providing a valued commodity (Cabanita maize) for subsistence and trade.

The Hualca-Hualca River picks up water from supplementary sources, including the Majes canal, as it winds it way down from the snow melt of the looming Hualca-Hualca Mountain. River water in Cabanaconde passes directly and continuously to the fields (figure 5.3), and unlike other Colca Valley communities, nocturnal irrigation is practiced.[9] Reservoirs, used for storing water at night in other communities, are not a feature of the system here. Annual rainfall is extremely variable in Cabanaconde, and periodic drought occurs in this area of the Andes.[10]

Although the heads of households and the water authorities are predominantly men, both men and women are skilled and active in irrigation from a young age. There were some 865 irrigators (usually heads of households) in Cabanaconde in 1980, according to the Ministry of Agriculture. They are organized in a water-users association, the Irrigators Commission, which is made up of local community members and which decides the irrigation cycle. The irrigation cycle determines the planting schedule and other agricultural activities, and is an important arena of social interaction. Family and interfamily ties and alliances, as well as disputes over land, are often expressed in water use — one waters land that one intends to seed. In these conflicts water can be used as a weapon: A slight turn of a rock can send water streaming into the already irrigated plot of an enemy, destroying the small plants.

Scratches and Gouges, Community and State

Power is inscribed in Cabanaconde's landscape in the form of water supply systems. In some places the inscriptions are fresh. Others are well routed from years of heavy use, and still others have been almost entirely erased. While the Peruvian state began to intervene in the irrigation practices of the community as early as the 1940s, it is only during the last twenty-five years that state influence in local water control has grown considerably. Yet since at least 1916, communal attempts have been made to secure a greater supply of water in order to expand the area of cultivated land. Some of these efforts have been successful; many others have not. The principal means of expanding the annual availability of water have been the cleaning of the high springs that feed the Hualca-Hualca River, through the maintenance and rehabilitation of canals built in pre–Columbian times, and through seeking out new sources of water. The latter includes the attempts to gain water from the contested area of Huataq and the opening of the Majes canal in 1983.

The ancient canal of Huataq represents a massive undertaking, surely the work of the organizing force of the Inka Empire. The Majes canal involved not only the modern Peruvian state, but an international consortium. At the other end of the scale, small scratches next to the large gouges of Huataq and Majes, are the very small springs and channels of water at the high end of the Hualca-Hualca River Basin that are dug out every year to increase the flow of water. In between these two extremes there are the heavily used canals, intakes, and secondary canals of the fields themselves. Although these canals receive constant minor and occasional larger modifications, they have been in continuous use for at least five hundred years. Here, too, power is inscribed in the different culturally determined forms of distribution that guide their use today. Let us examine each one of these inscriptions.

Cabaneño Ethnohydrology and the Local Model of Irrigation

The first of these hydraulic domains concerns the cultural construction of water and power that structures, in large part, the local form of water distribution. It is impossible to overstate the symbolic importance of irrigation for the Cabaneños—it pervades their lives, and is cause for great joy and great suffering. The multitude of irrigation-related offerings to Hualca-Hualca Mountain, the festivities and rituals associated with the irrigation rounds (*yakutinkay*), and the solemnness with which I was repeatedly told "we owe our lives to the water" ("el agua es nuestra vida") all convey the symbolic and emotional centrality of irrigation water. The Cabaneños are also proficient hydrologists who possess a vast store of technical knowledge concerning water flows, subterranean filtration, canal and terrace construction, and the changing chemical properties of water at different times of the year. This knowledge, combined with religious ideas concerning the correct ritual formulas for fertility, abundance of water, and personal safety is what I call "Cabaneño ethnohydrology," a variant of "Andean ethnohydrology" (see Sherbondy 1982).

The cult of Hualca-Hualca Mountain, a principal deity and the object of much worship and ritual for the people of Cabanaconde, is ancient. In 1586 the ethnic lords of the ancient Cavana polity told a Spanish crown official that their ancestors emerged from Hualca-Hualca Mountain, the source of their irrigation water. The same document states that the Cavana people dutifully worshipped her (Ulloa Mogollón [1586] 1965). Beliefs about origins in, and worship of, mountains and other sources of water is an ancient and widespread characteristic of many ethnic groups in the Andes (Sherbondy 1982; Reinhard 1985). Today, the irrigation and ritual practices associated with the cult of Hualca-Hualca Mountain are exemplified in what I have elsewhere called the "local model" of irrigation (Gelles 1994).

During most of the yearly distribution cycle, water management is in the hands of two men who carry snake-headed staffs of authority and often have flowers adorning their hats. They alternate, each spending four consecutive days and nights in the fields "together with the water." These are the water mayors (*yaku alcaldes*), the men responsible for the distribution of irrigation water. They are at the center of the local model of dual organization and distribute water to the lands classified as belonging to their moiety, either "*anansaya*" or "*urinsaya*."[11]

The ritual attainment of fertility and the symbolic control over the sources of water through worship of mountain gods and the earth are crucial aspects of the local model, and its rationale and practice must be understood in terms of these. Water comes from Hualca-Hualca Mountain and is therefore sacred. Water, however, can also bewitch, and can even be deadly to those who do not carry out the proper rituals. An overriding conceptual dualism that seeks equilibrium through the complementarity of opposites is another fundamental part of this cosmology. Historical evidence demonstrates the great time depth of these different conceptual domains. They continue to inform many other activities besides irrigation, such as herding and health care, as well as communal, family, and gender ideologies.

Irrigation structures the private and collective calendars of the Cabaneños during much of the year. It gives life to their terraced fields and to many of their conversations, and—as evidenced in the centuries-old cult of Hualca-Hualca Mountain—water is a basic part of their ethnic and communal identities. Mountains are gendered, and for the Cabaneños it is generally female mountains who provide water. Indeed, Hualca-Hualca is today referred to as the "mother" of the town. In Cabanaconde, unlike in other highland regions where people conceptualize irrigation water as semen or blood, water is likened to mother's milk, another essential, life-giving fluid. Cabanaconde also differs in that it has no "mother canal" or "mother reservoir" that is ritually cleaned by all of the irrigators as in the "Water Fiesta" described for many other parts of the highlands, and even other Colca communities.[12] Unlike the single eight-day communal catharsis associated with the spiritual and physical cleansing of the canals in the Water Fiesta, Cabaneño irrigation rituals are carried out by the water mayors and the irrigators throughout the distribution cycle. Previous to the advent of the Majes water in 1983, however, there was a ritualized communal work project that in some ways resembled the Water Fiesta described for other communities.

The Journey to Hualca-Hualca

Until 1983, the Journey to Hualca-Hualca (la Salida a Hualca-Hualca) was an annual event involving the entire community for the explicit purpose of increasing water availability for irrigation. This was accomplished by physically cleaning the small tributaries to Hualca-Hualca River. Availability was also enhanced by the ritual appeasement of the mountain through individual offerings. Because of its ritual nature, the "salida" resembled pilgrimages made to sacred mountains by other communities in the southern Andes (Sallnow 1987). In Cabanaconde the purpose of these rituals was to honor the mother mountain of the community, create an abundance of water, and secure the personal safety of the irrigators.

Communal authorities decided the date of the expedition, which would happen between the months of September and November. It often occurred twice a year, especially during drought years. "It used to look like an army," as one man put it. The townspeople, organized into their respective quarters (*cuarteles*), would march with mules and burros over 15 kilometers and up 2000 vertical meters (see figure 5.1) to clean the small rivulets or water channels (*sanqas*) that emerge from the high slopes and furrow down through the boggy irrigated pastures (*bofedales*) to the Hualca-Hualca River. Although the first record I have of these expeditions is 1922 (BMC), this annual pilgrimage is most likely an ancient feature of the ritual and irrigation complex associated with Hualca-Hualca Mountain.

A good part of the community used to make this march, remaining at the foot of Hualca-Hualca for three days. Not only would the participants dig out the small furrows on the lower slopes, but they would also climb to the snow of the mountain itself at around 5,500 meters. Here they had to cut a central channel through the snow, as well as other smaller furrows to redirect the snowmelt to the main channel, thereby increasing the overall volume of water. The work was strenuous. "You'd hit the snow three times with your pick, and you'd be out of breath," as one informant put it. Many people were known to faint, become ill, and even die. Indeed, in November 1934, a state representative who arrived to "organize the irrigators of the district" was told by the communal authorities that "the small flow of water to irrigate the fields each year is due to the efforts of this Town's inhabitants, who, making great sacrifices on the snow-covered mountain Hualca-Hualca, toil in the snow for eight days and at times even longer. It is well known that all who go there become snow-blind and gravely ill, many heads of family having passed away in the sacrifice to increase the waters" (BMC).

Ecologically and ritually, this annual cleaning is very different from the Water Fiesta of other communities. First, in Cabanaconde the collective labor of the community was mobilized to secure and increase supplementary sources of water by the opening of the sanqas, rather than to clean a major canal. The ritual aspects of the journey also differed from those found in the Water Fiesta of other regions in that major rituals were not carried out by communal authorities, but rather each family prepared its own offerings, leaving them in the snow or at the foot of the mother mountain. The Journey to Hualca-Hualca has not been made since 1983, when the water from the Majes canal was secured by the community. Many people with whom I spoke lamented the water lost by not going to Hualca-Hualca, also noting that these cleanings used to build solidarity among the townspeople, solidarity that spilled over into other communal venues. Ironically, the courageous effort to open the Majes canal (see later discussion), one that required great communal unity, brought to an end the salida—part work party, part pilgrimage—and with it an annually renewed solidarity.

The Ancient Canal of Huataq

In addition to these communal efforts, conflicts between the community and regional forces have great impact on the availability of water. An example is the case of Huataq, a spring with over 600 liters per second that exists in the high pastures (at about 4,500 meters) of the community (figure 5.1). Since at least 1916, the community has attempted to rebuild through communal work projects the 35 kilometers of canal that, during Inka times, apparently brought water to what were then cultivated lands.[13] In the early 1980s when this project was renewed in earnest with every sign of success, political manipulation by powerful individuals in Arequipa—including Ministry of Agriculture officials related to landowners in the neighboring community of Lluta and further downriver who also use Huataq's waters—brought the project to a halt.

Competition between communities over highland sources of water is often intense, and was also an important factor here. The neighboring community of Lluta, which also uses the Huataq waters, has fought legal and even physical battles with Cabanaconde over these waters since at least 1933. In that year there was a trial over Huataq, and in 1937 the entire town of Cabanaconde went "to take possession" (BMC). Attempts by the Cabaneños to repair the ancient canal throughout the 1930s, 1940s, and 1950s met with continued opposition from Lluta. The struggle became especially intense in 1968. Each community had National Decrees to demonstrate their lawful possession of the Huataq waters, and each accused the other of trying to steal the water. Cabanaconde pointed to the pre–Columbian canals of Huataq, "which have been used by Cabanaconde since time immemorial . . . at present the liquid element remains in the pastoral lands of Cabanaconde" (BIC). The mayors of Lluta and Cabanaconde each presented the subprefect of the province with documents asserting their rights to the water. The waters of Huataq were then legally divided into three parts, two for Cabanaconde and one for Lluta. It was agreed that Lluta would not make another canal or attempt to gather more water from Huataq until Cabanaconde finished a canal to take water from Huataq to the town's fields.

The conflict, however, did not die there. In September 1969 Cabanaconde denounced the people of Lluta "for the abuses committed to the waters of Huataq" (BMC). In 1971 a contradictory resolution appeared that stated that the waters of Huataq belong to Lluta. In December 1979 the legal battle had shifted again, this time in Cabanaconde's favor: "the transfer of the Huataq waters . . . was favorable" (BMC). Other resolutions soon appeared, however, and the ultimate status of the Huataq waters has yet to be decided. In 1988 Lluta was still threatening violence should the Cabaneños renew their efforts to divert this water.

The numerous and contradictory national decrees about Huataq demonstrate the capricious nature of state intervention in highland water politics. And in this case, some of the most powerful representatives of the state in the Ministry of Agriculture were in fact beholden to regional elites with a personal stake in the Huataq waters. These obstacles did not, however, diminish the Cabaneños' resolve. As late as 1985, and even after having gained access to the waters of the Majes canal, the community continued to send engineers and communal work parties to rehabilitate the ancient canal of Huataq. Because of political influence, the conflict with Lluta, and the fact that this project clearly demands the logistical support of the Peruvian state, the Cabaneños never completed this difficult job. The community did prevail in a different case, however, and when it did, the Huataq waters became an important bargaining chip.

The Majes Canal

Another source of water, as well as of contention, is the large canal built by the Majes Consortium in the late 1970s. This canal, which pipes highland water to the arid coast, passes through communal territory. As figure 5.4 shows, the map elaborated by this billion-dollar development project neglected to show that in the path of the proposed canal lay more than a dozen communities and tens of thousands of peasants. This is symptomatic of the regard that Majes and the Peruvian state had for the inhabitants of Cabanaconde and the Colca Valley.

The role of the state is clear. As early as May 6, 1967, the Ministry of Development and Public Works by way of the Arequipa Board of Rehabilitation and Development stated that Cabanaconde "is being considered for three thousand hectares" to favor the irrigators there (BIC).[14] Promises were made but not kept. Until 1983 there were no benefits from the project, except an improved road and poorly paid, temporary, and dangerous jobs. The project instead brought widespread social and environmental problems for the communities of this region (Hurley 1978; Benavides 1983; Sven 1986).

Workers from the project, housed in a large encampment near the community, abused the local townsfolk, treating them as so much infrastructure. Many workers would not buy products from women accompanied by their husbands, and there were incidents of prostitution and rape. Although the improved road provided the means for greater mobility up and down the valley and to and from Arequipa, the community was also subjected to economic, cultural, and political forces it had never known before. Large amounts of money began to flow into the community, and many new small stores were opened to meet the needs of a boom econ-

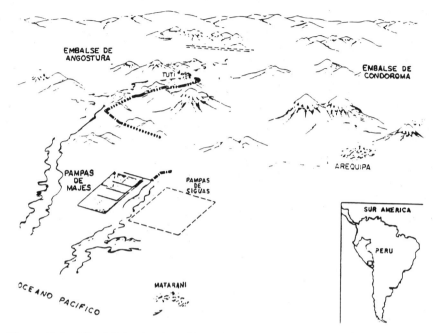

Figure 5.4 The Majes Project, Peru (source: Autodema, n.d.).

omy. Most Cabaneños with whom I discussed the Majes Project lamented the changes and abuses the project brought. "Everything became money, money, money," as one man put it.

Harder to express are the profound social and cultural changes the community experienced, including the way local society defined itself. "Criollo" views, disparaging towards the "simple" and "backwards ways" of the highland dwellers, became widely felt. As the president of the community ruefully expressed, "the workers would come rolling into town. They were from all over—Cuzco, Puno, the coast—all over. They'd come in and say in Spanish 'son of a bitch, it's hot! Hey, give me a case of beer,' and pretty soon all the boys in town were walking around, saying 'hey, son of a bitch.'" Less respect for elders, an increase in vandalism, and the breaking down of social mores were other correlates of Majes. The local culture was denigrated, and several rituals, such as the *torotinkay* during the sowing, disappeared. The workers even instituted a new saint in the community.[15]

The social impact was paralleled by an ecological one. Economic dependence on income generated by the project increased as the community's resources suffered. During a series of drought years, when the volume of river water was already extremely low, the project used large amounts of water from the river for its operations without the community's permission (BIC). Canals and terraces were damaged by project roads. The underground tunnels built by the project also affected subterranean sources of water.[16] Because of the drought and Majes' insensitive use of the little river water that remained, the cultivated area decreased dramatically. A petition sent by the Irrigators Commission to the Ministry of Agriculture on March 18, 1980, states that the devastation caused by "the last droughts" and the water supplied to Majes has had "horrific results . . . our agricultural fields have diminished by 80 percent."[17]

Abuses were tolerated, in part, because of the irrigation water and extended land base the community expected to receive from the project. The project also promised the community to use its engineers and heavy machinery to improve the drainage of the Hualca-Hualca River Basin and to recover the canal of Huataq (BIC). Through 1979 and 1980 the Majes Project continued to promise support for these projects as well as enough water, 1000 liters per second, to recover several thousand hectares for the community.[18] But it soon became clear to the community that Majes had no intention of carrying through with their promises. The first hint of resistance became manifest.

In March 1980, a commission made up of the president of the Peasant Community, the president of the Irrigators Commission, and the mayor of Cabanaconde again requested water from the Majes canal in the form of a sluice gate (*ventana*) and assistance in improving the waters of Hualca-Hualca. They stated that "the District of Cabanaconde has been forgotten," and that "everyone will unite as one man" if their demands went unanswered. Later that month, a memorandum was sent out to the Ministry of Agriculture and to the president of the republic, "clamoring for one thing alone, which is water" (BMC). In August 1981, a letter to the Majes Consortium states that if it does not recompense the community for its water, "the agriculturalists will stop the services they provide . . . the higher authorities are fooling us." More ominously, the letter continues, "the townspeople will take action to the last consequences in the case that an immediate solution is not arrived at." Yet no water was allotted to the community. In September 1981, the books of the Irrigators Commission refer to the drought of that year as "a frightful crisis."

In January 1983, another letter was sent by the mayor of Cabanaconde to the president of Peru, Fernando Belaúnde Terry, stating that "Cabanaconde is the most populated district of the Province of Caylloma. . . . The ratio of man/land is unequal, and all community members are small holders, a situation which generates the massive exodus of the population . . . the actual [water] capacity is only 80 liters per second, insufficient for 1200 hectares. We irrigate every 100 days. . . . This generates poverty, undernourishment, infant and adult mortality, desertion from school, alcoholism and illiteracy. . . . Cabanaconde has been marginalized by the Majes Macon Project, whose canal transverses our jurisdiction and has destroyed our natural resources, such as land and water, making even graver the already precarious economic situation of this community."

They asked President Belaúnde to authorize a sluice gate from the Majes canal; money, machines, and technical assistance to reconstruct the Huataq canal; the construction of a cement reservoir in Joyas; and the settlement of families in the Pampas de Majes (the target area of the project). The letter was signed by the communal authorities and several hundred people. A few days later, a memo was sent to the Ministry of Agriculture asking "for a sluice gate in the Tomanta Sector," which is where the Majes canal crosses the Hualca-Hualca River. These letters, as well as the many and various pleas the community had made to the consortium and to the regional authorities, fell on deaf ears. While the remaining plants withered under the intense Andean sun in the most serious drought of the last thirty years, thousands of liters of water per second were streaming by the community, sequestered in a thick cement canal. The possibility of famine became real.

In March 1983, the Cabaneños opened the Majes canal in a classic case of peasant resistance. The now renowned "eleven heroes" of the community, some of whom were authorities of the Irrigators Commission and Peasant Community, went nightly to drill a hole in the thick cement casing of the canal where it crossed the Hualca-Hualca River (see figure 5.4).[19] Finally, dynamite was used. People soon began to comment that the volume of the Hualca-Hualca River had increased, and an assembly was hurriedly called. The entire community swung into action.

A permanent guard kept watch in the church tower, ready to sound the bell should the police arrive. A trumpeter was stationed at the entrance of the town, and barricades were built on the road. The eleven heroes left the community or slept in the orchards deep in the Colca Valley. A committee traveled to Arequipa to report what had been done, and they were immediately arrested. But the Cabaneños had skillfully published several news clips in Arequipa newspapers in the weeks before they opened the canal, decrying the drought and the way that Majes had lied to the community, as well as the lack of government support. This was done not only to assert the Cabaneños' rights to the water, but to insure that their actions not be confused or purposefully misinterpreted as those of terrorists.

A police contingent was sent to the community, but when they arrived the entire community confronted them. The community claimed collective responsibility for the opening of the canal, and demanded that the water not be withdrawn. Several large machines of the consortium were taken hostage. A few days later, the subprefect of the region, the mayor of Cabanaconde, and other important authorities were present in the plaza of the community to talk "about the problem of the window opened by the townspeople in the Tomanta Sector" (BIC). The mayor exclaimed, "the Ministry of Agriculture is guilty because it has never taken any steps to help the agriculture of Cabanaconde." The subprefect agreed to provide a legal transfer of the waters within the briefest time possible, promising that there would "be no repression against the townspeople." But when he asked that the machines be returned to the Majes consortium, his request was denied by the community. The mayor demanded that 300 liters per second be given to the community. After he agreed to the conditions set by the community, the subprefect received an ovation. People still talk excitedly and proudly about how the entire community took responsibility, and how they were ready to fight to the end for the water.

The communal authorities, with help from the migrant associations of the community in Arequipa and Lima, continued to negotiate with the regional authorities, explaining that they were not terrorists, but were dying of hunger. Out of fear of further conflicts, Autodema (Autoridad Autónoma de Majes) — the administrative unit of the Majes canal — finally agreed to cede 150 liters per second to Cabanaconde. This resulted in the Cabaneños becoming heroes for the entire valley. In August 1983, the hole was patched and a valve was installed at Tomanta. The next day the entire community went to Tomanta in procession, with a band at its head. "After the blessing by the priest the two valves were opened in the presence of the president of the Irrigators Commission and the townspeople" (BIC). The Cabaneños' heroic feat set an example for the other communities of the left bank of the Colca Valley. They threatened to take similar action "or call in the boys from Cabanaconde," as several Cabaneños proudly told me. These other communities were soon given access to the "Majes water."

Once the flow of 150 liters per second from Majes was secured in 1983, attempts to increase the area of cultivated land began almost immediately. Canals in the lower part of the agricultural fields were extended by means of communal labor, and a large number of abandoned terraced fields were rehabilitated. By 1988, several dozen hectares of newly recovered fields were producing good harvests. In 1989, the community negotiated an additional 200 liters per second from the Majes Project Administration and the Ministry of Agriculture by signing a document promising not to touch the Huataq waters in the future. By August 1994, several hundred hectares of terraced fields — abandoned after, and not cultivated since, the Spanish invasion — were yielding bountiful harvests. And although this land recovery has been anything but a trouble-free process (see Gelles 1994), the Cabaneños' courageous actions and wily maneuvering have provided them with a relatively large and growing land base.[20]

Cabanaconde, "Development," and the State

Despite the fact that Cabanaconde has now increased its area of cultivated fields, initiatives taken by the Peruvian state clearly had little to do with this. State representatives consistently ignored the community's cries for assistance, and in the case of the Majes canal, actively worked against community interests. During the time of Majes and the terrible drought of the early 1980s, the state added insult to injury by instituting a water tariff. Among the demands that regional and state authorities had to accede to when the Cabaneños opened the Majes canal was that the tariff be rescinded. Cabanaconde has not paid the tariff since then.

The case of the Majes Project, and the resistance of the Cabaneños to it, is exceptional. The neglect and lack of respect the state has shown for Andean communities and their irrigation systems is widespread. This involves the cultural politics of the coast and the highlands in Peru, in which the relationship between the two areas is completely infused with a colonial mindset that views the human and natural resources of the Andean highlands as inferior to the *criollo* coast. This is expressed in a Ministry of Agriculture document elaborated in 1980, which states that the problem of Colca Valley villages (and surely the highlands as a whole) is due to "the low cultural level of the irrigators . . . [and] a certain resistance of the agriculturalist to rational work methods" (ORDEA 1980). Yet during this same year, the water tariff was being calculated, and Cabanaconde's water was being usurped by Majes.

A related form of cultural hegemony is found in the state-imposed model of water distribution. Since the turn of this century, the Peruvian state has declared that all water in Peru is the property of the state (Andaluz and Valdez 1987). It is only since the 1960s, however, that the state has centered its energies on extending its control to highland irrigation systems by imposing water user associations and new forms of distribution. This is what Lynch (1988) has called the "bureaucratic transition."[21] The Peruvian state instituted irrigation districts, sectors, and water-user associations based on Spanish and more contemporary bureaucratic traditions similar to state-financed systems throughout the world. In many Andean communities, including Cabanaconde, however, the bureaucratic transition has yet to take place (Gelles 1994, 1995).

In Cabanaconde, resistance to state encroachment in water management is evident in the persistence of the office of water mayor and the local model of distribution described earlier. Since the 1970s, the state has consistently attempted to intervene in the distribution practices of the community. Today, for a brief but important period during the yearly agricultural cycle, there is a different type of "repartitioner," or person in charge of distribution. These individuals, who are community members like the water mayors, implement the state's model of distribution. They are called controllers (*controladores*), and distribute water to the fields sequentially from "one end to the other," ignoring the dual classification of the plots (anansaya and urinsaya) found in the local model. A whole series of cultural norms accompany the state model of distribution, and they are different from those of the local model. The controllers, for example, do not receive coca and liquor from the irrigators as the water mayors do. Rather, they receive money. They are not fulfilling a civil office (*cargo*), as are the water mayors, but rather a minor civic duty. Instead of a snake-headed staff to legitimate their authority, the controllers have an official decree from the local Irrigators Commission. They neither perform elaborate rituals, nor sponsor large social events, as do the water mayors.

State and local models of irrigation represent two distinct ways of conceptualizing and implementing water management. One takes a secular and bureaucratic view of water management, while the other is focused on ritual assurance, and views water as part of a larger social and symbolic universe. Today, and over the last fifty years, there have been attempts to completely supplant the local model of distribution for that of the state model. Although the state model has gained ground over the years, as of 1994 the local model was still firmly entrenched. Adherence to the local model of irrigation in Cabanaconde and elsewhere in the Andes must be understood in the larger context of the cultural politics of coast and highlands. Andean peasants are dominated peoples within a nation-state that does not share, nor have respect for, their cultural values. These attitudes often find institutional and violent expression.

But Andean peoples fight back, and there is a growing literature on cultural resistance and ethnic identity in the Andes. In his study of power and authority among the Yura (a Bolivian highland society), Rasnake (1988) found that traditional staffbearers are linked through ritual action to fundamental cosmological forces, and that this office serves to recreate ethnic identity. Traditional authorities and the rituals they perform legitimate "the broader institutional framework of the Yura ethnic group and its relation to the encompassing society . . . [one that] refuses to acknowledge the validity of their own self-definition" (Rasnake 1988:269). Clearly, this same dynamic is at work in Cabanaconde, where the state tries to impose its meanings and world view through water-use systems (and many other means), and where the snake-headed staffs provide an especially powerful symbol of local meaning and ethnic identity.

Resistance to state interference must also be understood in terms of the political ecology of irrigation development. As detailed throughout this chapter, the "availability" of water is largely determined by cultural and political forces at local and supra-local levels. The Journey to Hualca-Hualca, in which the community deployed its forces to increase the quantity of river water, was intimately linked to certain principles of Cabaneño ethnohydrology, and to the cosmological referents of the local model of water distribution. The brief case study of the Huataq canal revealed some of the ways that regional elites and state actors, as well as intercommunity politics, impact the local availability of water. Finally, the construction of the Majes canal through the Colca Valley, which was done to support large-scale coastal agriculture and help service Peru's foreign debt, showed how local irrigation development is also conditioned by national and international forces.

The billion-dollar Majes Project steamrollered the communities of this area socially, economically, environmentally, and culturally. With relatively little expense (as exemplified by the minimal costs of diverting water to Cabanaconde), the project could have included the Colca communities in their plans, helping to expand the valley's cultivable area. If these communities have benefited from the project to date, it is due to their own courageous initiative. Indeed, the diversion of the Majes water—as well as the more common "everyday forms of peasant resistance" (Scott 1985) such as the non-elaboration of a water-users list, the nonpayment of tariffs, and the persistence of the local model of irrigation—provide poignant examples of ways that local communities struggle with and frustrate intrusive state initiatives.

Historically and culturally informed analyses of local irrigation systems (and other forms of common property) have great potential, both in terms of theory, and as an applied research tool. Indeed, the approach I have presented here can be extended to other communities, development projects, and national contexts. In Latin America, for example, such an approach would seem especially applicable to Guatemala, Mexico, Bolivia, and Ecuador. In these countries, indigenous people today constitute a large percentage of the population (in some cases a cultural majority), and part of the extensive farming systems established by the expansive indigenous states that flourished in pre–Columbian times continues to be used. As their growing populations put more and more pressure on the land base, many communities are actively attempting to rehabilitate abandoned cultivated fields and the indigenous technologies—canals, reservoirs, raised fields, terraces, and so on—that sustain them.

Yet modern nation-states, and irrigation development initiatives in general, do not invest in such technologies, nor in the cultural models of those who manage them. Rather, the world views encoding the implementation of these technologies are excluded from, and denigrated by, national development discourses. International irrigation development initiatives do the same, putting more faith in abstract and supposedly cross-culturally valid formulas than in contextualizing analyses. One result of this is that a good many irrigation projects, implemented in a top-down manner unresponsive to local cultural and political considerations, like the Majes canal, have enormously adverse effects on rural populations. Clearly, more attention needs to be paid to indigenous models of irrigation, and to the ways in which local understandings of "development" condition state and community relations. Irrigation research needs to move away from mechanistic views of society to ones that evoke the power politics that generally lie at the heart of irrigation projects and their problems.

Acknowledgments

The material presented here is based on fieldwork carried out in the central Peruvian Andes between 1982 and 1984, and in the southern Peruvian Andes between 1987 and 1988. The Tinker Foundation funded predissertation summer research in 1985. The Fulbright-Hays Commission, the Social Science Research Council, and the Interamerican Foundation provided financial support for the 1987–1988 field season. During the different periods of research I was affiliated with the Anthropology Department of the Pontificia Universidad Católica of Lima. I would like to express my thanks to all of these institutions, and to the Ciriacy-Wantrup Post-Doctoral Fellowship of the University of California at Berkeley. Many thanks go to Soledad Gelles, Emery Roe, Catherine Macklin, Seth Macinko, Nancy Peluso, Ann Hawkins, Sara Michaels, Louise Fortmann, and Vinay Gidwani for their comments on an earlier draft. The maps were drawn originally by Nancy Lambert-Brown, except for figure 5.4, which is from Autodema (n.d.). All of the translations from Spanish and Quechua are mine, except where noted.

Notes

1. My usage borrows from those who view political ecology as combining the concerns of ecology and political economy (e.g., Blakie and Brookfield 1987; Schmink and Wood 1992; Sheridan 1988; Peluso 1992), but differs in that it places emphasis on the way that politics and natural resources are culturally modeled (see also Gelles 1993a). This is part of a larger shift in irrigation studies (e.g., Geertz 1980; Lansing 1991; Sherbondy 1994). In other places I have detailed the historical roots and cultural models found in Cabanaconde's irrigation system (Gelles 1994, 1995). In the present essay, I focus on the development of the community's hydrological resources.

2. State intervention in local irrigation practices has also had positive effects, for example, helping democratize certain political processes in the community. In this essay I am only concerned with showing the state's role, largely negative, in developing the hydraulic resources of the community.

3. The very elaborate female dress and the hats for both men and women clearly mark ethnic boundaries between Cabanaconde and other Colca Valley communities. Beyond this very visible marker, there are differences in many practices such as funeral rituals, sowing rituals, weaving, sewing, and to a certain degree, Quechua pronunciation and vocabulary (see Femenias 1987).

4. Over 27,000 people, a good part of them Andean peasants, have thus far been killed in the dirty war; many highland communities have been completely decimated. Although the region surrounding Cabanaconde was at one time declared a "red zone," the town has for the most part stood outside the crossfire between Shining Path and the military forces.

5. Before and during the Inka Empire, Cabanaconde "was more important politically

and in productive terms than other zones of the Colca Valley" (de la Vera Cruz 1987:10). Irrigation has been linked to power in this area since pre–Inka times, and state systems have seized on this hydrological system and the people dependent upon it since time immemorial. The Inka imposed different social and administrative divisions in Cabanaconde that continued to structure social relations well after the Spanish conquest.

6. Despite being one of the most productive *encomiendas* during the early colonial period, diseases, civil wars, over-exploitation, and the opening of the mines of Caylloma in 1627 led to a tremendous demographic loss in Cabanaconde, reaching an extreme in the late seventeenth century. From the 1570s—when the dispersed villages of the immediate area were brought together to form the town of Cabanaconde through the Spanish colonial forced resettlement program called *reducción*—until the 1680s, the tributary population declined from 1345 to 256 (Cook 1982:17, 25).

7. Although it is difficult to get a precise figure for this shifting population, there were at least 600 households in the community as of 1988. Population growth has placed more and more pressure on land, water, and the other productive resources of the community. Although Cabanaconde "has by far the most land in cultivation and the most land in grains" (Denevan 1987:21) of the Colca Valley villages, it also has the largest population. The population pressure, combined with a bilateral inheritance pattern, creates rampant fractioning of land holdings. Because of this, and because Cabanaconde is a differentiated community, many farmers must often rely on sharecropping arrangements and land rental to make ends meet.

8. Because of limited space, this brief characterization of Cabanaconde must suffice. Obviously, it cannot do justice to the complexity of the community and its links to national and international political and economic forces. I have explored a piece of this larger context elsewhere (Gelles 1993b).

9. The river water travels over 15 kilometers and more than 2500 vertical meters to the fields below. In fact, the Hualca-Hualca River could be considered a stream during most of the year. In the rainy season, river volume can reach over 1500 liters per second during periods of intense precipitation (Abril Benavides 1979:15). Because it is considered a river by the Cabaneños and maps of the region, I will retain the word "river." In rain-abundant years, a few farmers try to dry-farm additional fields (usually fodder crops). In any case, these dry-farmed fields (*terrenos de secano*), which constitute well under 1 percent of the total cultivated area, usually receive irrigation water at some point in their growing cycle. The absence of reservoirs in Cabanaconde distinguishes it from many of the upper valley communities such as Coporaque (Treacy 1989, 1994), Lari (Guillet 1987), and Yanque (Valderrama and Escalante 1988), as well as nearby Tapay (Paerregaard 1994).

10. The Hualca-Hualca River supplies Cabanaconde's agriculture with between an estimated 75 and 150 liters per second of water during the dry season months of May to December (Abril Benavides 1979; ORDEA 1980). Before the water of the Majes canal became available in 1983, each complete cycle or round of irrigation water through all the cultivated fields lasted 90 to 120 days. The opening of the Majes canal doubled the amount of available water, and today each round lasts 45 to 50 days. The additional water has intensified existing agriculture and has allowed Cabanaconde to expand its cultivated land. There is still, however, only one harvest a year. Although heavy rainfall is important for the

proper maturation of maize, precipitation also increases the snowpack on Hualca-Hualca and the volume of melt. Irrigation extends the growing season, permitting sowing four months before the rainy season and the plants' maturation before frost sets in, usually in May or June. This allows farmers to raise the upper limits of certain crops, such as maize (Mitchell 1981; Mayer 1985).

11. As he moves through the fields, there are several plots of land to which the water mayor will not give water and which must wait for the water mayor of the "opposing" side or moiety. Elsewhere, I have demonstrated that the model of dual organization, of which the water mayors are the centerpiece, is a legacy of the Inka and colonial Spanish states. Today, however, it constitutes the local "indigenous" form of irrigation management, and is related to a wide range of social and semantic dualisms found in the community (Gelles 1995).

12. These differences are quite notable. In Cabanaconde, as seen here, irrigation water is conceived of as feminine (*yakumama*) because of its association with Hualca-Hualca, the mother mountain and female deity, from which it descends. In many communities, including some in the upper Colca Valley (see, e.g., Valderrama and Escalante 1988; Treacy 1994), irrigation water is conceptualized as a vital life-giving, male, force—blood or semen—that emanates from the subterranean arteries and veins of the mountain deities to impregnate the "Earthmother" (see, e.g., Arguedas 1985; Isbell 1974; Ossio 1976; Gelles 1986; Valderrama and Escalante 1988). The large, communally sponsored, ritualized canal cleanings in other highland communities are known as the "Water Fiesta" or "Canal Scraping" (*Fiesta de Agua* or *Champería* in Spanish, *Yarqa Asp'iy* in Quechua).

13. Although it may be that the Cabana polity, and possibly Wari before it, made attempts to tap this source of water, the Huataq canal appears to have been built by the Inkas. In June and October of 1916, the people of Cabanaconde were trying to recover lands by rebuilding the Huataq canal, and the Municipality was providing some of the expenses such as sugar, trago, and coca (B M C). Work continued in October and December of 1919, and again in 1920 and 1921. The same source mentions an engineer inspecting Huataq in 1926, and one making a "scientific investigation" of the canal in 1932 (B M C). The very name "Huataq" (*wataq*), which means "tied down" in Quechua, suggests the difficulty the community has had with this water source.

14. Various other government agencies convened later that year. An entry in the *Books of the Irrigators Commission* states that the community is "[s]oliciting some three thousand liters of water from the Main Canal of Majes to be used in the irrigation of the fields now being cultivated and in the expansion of new lands." This same entry, dated November 10, 1966, reports that the Ministry of Development and Public Works, the Board of Rehabilitation of Arequipa, the National Fund for Economic Development, and the National Office of the Agrarian Reform were going to study the springs of the entire area so that if the canal of Majes affected them, Cabanaconde could claim damages (B I C).

15. The new saint was the Virgin of Chapi. On May 4, 1981, they moved the image of this virgin from the camp at Castropampa to Antesana, a hill above town. Today this saint has a dedicated following among *misti* schoolteachers and some community members.

16. On October 18, 1977, the Irrigators Commission complained that Majes project "is using water of Cabanaconde without authorization or permission, and there is a consider-

able diminution in the volume of water" (BIC). Canals were also severely damaged. The subterranean tunnels built by Majes apparently diverted some of the underground sources of water that fed the springs of the orchards—in some cases these dried up completely.

17. Official sources assert that, for 1979 and 1980, the cultivated area was 10 percent below the average for the period 1975 to 1981, and 25 percent smaller than the most productive of these years (see Denevan 1987:17). Many Cabaneños assured me that it was much worse.

18. In 1978, for example, the Majes Consortium pledged to send an engineer and help Cabanaconde obtain more water from Hualca-Hualca. Two years later they repeated their promise, stating that machines were available for improving the waters of Hualca-Hualca. They also promised assistance to recover the lost canal of Huataq, as well as water from the Majes canal "for three thousand hectares in Cabanaconde and Huambo." In 1979, 1980, and 1981, Majes again promised to help recover 1000 hectares of cultivable lands in Cabanaconde—it never carried through with its promises.

19. Today the eleven heroes meet in secret to celebrate that day in March 1983 when they opened the canal. When the agricultural fields were recently expanded, the heroes were given their choice over plots of land. Although everyone in the community knows who they are, they are still careful to not be identified by government officials for fear of reprisals.

20. The recovery of abandoned terraced fields and other indigenous technologies for cultivation, such as raised fields (see, e.g., Erickson and Candler 1989), is receiving increasing attention from scholars (see, e.g., Denevan 1987; Treacy 1989, 1994). Masson (1987:191) estimates that there are a million hectares of terraced fields in the Peruvian highlands, and that only 25 percent of these are currently in use. Cabanaconde's efforts to recover its lost infrastructure have been very successful. As of 1994, the community had recovered over 700 hectares; another 400 will be brought into production in the next few years. By then, the community will have essentially doubled its land base (Gelles 1994).

21. Lynch (1988, 1991, 1993) uses this term to describe the processes by which the Peruvian government has increasingly intervened in small scale highland irrigation during the last twenty years. For other studies on this theme, see Seligmann (1995) and Guillet (1994). Cabanaconde has had a special-purpose irrigation association composed of community members, the Irrigators Commission, linked to the state since the 1940s. The relatively early adoption of this institution was generated by political conflict within the town, and has had ambivalent effects. It has sometimes stemmed the abuses of powerful peasants, and at other times has facilitated them. Though the Irrigators Commission is a state office, supposedly conforming to the norms of the state's irrigation bureaucracy, to a large degree it follows the dictates of the local model of irrigation.

References Cited

Abril Benavides, Dionicio Nilo
    1979   *Estudio hidrogeológico de la cuenca del río Hualca-Hualca.* B.A. thesis, Department of Geology, Universidad Nacional San Agustín de Arequipa, Peru.

Andaluz, Antonio, and Walter Valdez

1987   *Derecho ecológico peruano: inventario normativo.* Lima: Editorial Gredes.

Arguedas, José María

1985   Puquio: A Culture in Process of Change. In *Yawar Fiesta,* trans. Frances Hornine Barraclough, pp. 149–92. Austin: University of Texas Press.

Autodema (Autoridad Autónoma de Majes)

n.d.   *Esto es Majes . . . Un Sueño Hecho Realidad.* Arequipa: Autodema.

Benavides, María

1983   *Two Traditional Andean Peasant Communities under the Stress of Market Penetration: Yanque and Madrigal in the Colca Valley, Peru.* M.A. thesis, University of Texas, Austin.

BIC Books and documents of the Irrigators Commission, Cabanaconde.

Blakie, Piers M., and Harold Brookfield

1987   *Land Degradation and Society.* New York: Methuen.

BMC Books and documents of the Municipal Council, Cabanaconde.

Brush, Stephen

1977   *Mountain, Field, and Family.* Philadelphia: University of Pennsylvania Press.

Cook, David N.

1982   *The People of the Colca Valley: A Population Study.* Boulder, Colo.: Westview Press.

De la Vera Cruz, Pablo

1987   Cambios en los patrones de asentamiento y el uso y abandono de los andenes en Cabanaconde, Valle del Colca, Perú. In *Pre-Hispanic Agricultural Fields in the Andean Region,* ed. W. Denevan, K. Mathewson, and G. Knapp, 89–128. Oxford: B.A.R. International Series.

Denevan, William

1987   Terrace Abandonment in the Colca Valley, Peru. In *Pre-Hispanic Agricultural Fields in the Andean Region,* ed. W. Denevan, K. Mathewson, and G. Knapp, 1–44. Oxford: B.A.R. International Series.

Erickson, Clark, and Kay Candler

1989   Raised Fields and Sustainable Agriculture in the Lake Titicaca Basin of Peru. In *Fragile Lands of Latin America: Strategies for Sustainable Development,* ed. John Browder, 230–49. Boulder, Colo.: Westview Press.

Femenias, Blenda

1987   Regional Dress of the Colca Valley, Peru: A Dynamic Tradition. Paper read at the symposium "Costume as Communication," Brown University, Providence, R.I.

Fuenzalida, Fernando

1970   La matriz colonial. *Revista del Museo Nacional (Lima)* 35:92–123.

Geertz, Clifford

1980   *Negara: The Theatre State in Nineteenth Century Bali.* Princeton: Princeton University Press.

Gelles, Paul H.

1986   Sociedades hidraúlicas en los Andes: Algunas perspectivas desde Huarochirí. *Allpanchis Phuturinga, Cuzco* 18(27): 99–147.

1993a   Irrigation as a Cultural System: Introductory remarks. *Proceedings of the 24th Chacmool Conference*, 329–32. Alberta: University of Calgary Archeological Association.

1993b   *Transnational Fiesta: 1992, Teaching Notes.* Berkeley: Center for Media and Independent Learning.

1994   Channels of Power, Fields of Contention: The Politics of Irrigation and Land Recovery in an Andean Peasant Community. In *Irrigation at High Altitudes: The Social Organization of Water Control Systems in the Andes,* ed. W. P. Mitchell and D. Guillet, 233–73. Society for Latin American Anthropology Publication Series, Vol. 12. Washington, D.C.: American Anthropological Association.

1995   Equilibrium and Extraction: Dual Organization in the Andes. *American Ethnologist* 22(4):710–42.

Guillet, David

1987   Terracing and Irrigation in the Peruvian Highlands. *Current Anthropology* 28(4):409–30.

1994   Canal Irrigation and the State: The 1969 Water Law and Irrigation Systems of the Colca Valley of Southwestern Peru. In *Irrigation at High Altitudes: The Social Organization of Water Control Systems in the Andes,* eds. W. P. Mitchell and D. Guillet, 167–88. Society for Latin American Anthropology Publication Series, Vol. 12. Washington, D.C.: American Anthropological Association.

Hurley, William

1978   *Highland Peasants and Rural Development in Southern Peru: The Colca Valley and the Majes Project.* Ph.D. dissertation, Department of Anthropology, Oxford University, Oxford.

Isbell, Billy Jean

1974   Kuyaq: Those Who Love Me. An Analysis of Andean Kinship and Reciprocity Within a Ritual. In *Reciprocidad e intercambio en los Andes peruanos,* ed. G. Alberti and E. Mayer, 110–52. Lima: Instituto de Estudios Peruanos.

Lansing, J. Stephen

1991   *The Computer and the Goddess: Technologies of Power in the Engineered Landscape of Bali.* Princeton: Princeton University Press.

Lynch, Barbara

1988   *The Bureaucratic Transition: Peruvian Government Intervention in Sierra Small Scale Irrigation.* Ph.D. dissertation, Department of Rural Sociology, Cornell University, Ithaca.

1991   Women and Irrigation in Highland Peru. *Society and Natural Resources* 4:37–52.

1993   The Bureaucratic Tradition and Women's Invisibility in Irrigation. In *Proceedings of the 24th Chacmool Conference*, 333–42. Alberta: University of Calgary Archeological Association.

Masson, Luis

1987   La ocupación de andenes en Perú. *Pensamiento Iberoamericano* 12(2):179–200.

Mayer, Enrique
  1985   Production Zones. In *Andean Ecology and Civilization*, eds. S. Masuda, S. Shi-
          mada, and C. Morris, 45–84. Tokyo: University of Tokyo Press.
Ministry of Agriculture, Peru
  1987   *Diagnóstico de Cabanaconde*. Arequipa: Government Press.
  1980   *Relación de recibos y tarifa de agua para el Distrito de Cabanaconde*. Arequipa:
          Government Press.
Mitchell, William P.
  1981   La agricultura hidraúlica en los Andes: Implicaciones evolucionarias. *Tecnolo-
          gía del mundo andino*, Vol. 1, ed. H. Lechtman and A. Soldi, 45–167. Mexico
          D.F.: U.N.A.M.
ORDEA
  1980   *Diagnóstico del distrito de riego no. 49: Colca*. Arequipa: Sub-dirección Nacional
          de Aguas y Suelo.
Ossio, Juan M.
  1976   El simbolismo del agua en la representación del tiempo y el espacio en la fiesta
          de la acequia en Andamarca. Mimeograph, Pontificia Universidad Católica,
          Lima.
Paerregaard, Karsten
  1994   Why Fight over Water? Power, Conflicts, and Irrigation in an Andean Vil-
          lage. In *Irrigation at High Altitudes: The Social Organization of Water Control
          Systems in the Andes*, eds. W. P. Mitchell and D. Guillet, 189–202. Society for
          Latin American Anthropology Publication Series, Vol. 12. Washington, D.C.:
          American Anthropological Association.
Peluso, Nancy L.
  1992   *Rich Forests, Poor People. Resource Control and Resistance in Java*. Berkeley:
          University of California Press.
Rasnake, Roger
  1988   *Domination and Cultural Resistance: Authority and Power among an Andean
          People*. Durham: Duke University Press.
Reinhard, Joseph
  1985   Chavin and Tiahuanaco: A New Look at Two Andean Ceremonial Centers.
          National Geographic Research Reports 1(3):395–422.
Sallnow, Michael
  1987   *Pilgrims of the Andes: Regional Cults in Cusco*. Washington, D.C.: Smithsonian
          Institution Press.
Schmink, Marianne, and Charles D. Wood
  1992   *Contested Frontiers in Amazonia*. New York: Columbia University Press.
Scott, James C.
  1985   *Weapons of the Weak: Everyday Forms of Peasant Resistance*. New Haven: Yale
          University Press
Seligmann, Linda
  1995   *Between Reform and Revolution: Political Struggles in the Peruvian Andes, 1969–
          1991*. Stanford: Stanford University Press.

Sherbondy, Jeanette

1982   El regadío, los lagos y los mitos de origen. *Allpanchis (Cuzco)* 17(20):3–32.

1994   Water and Power: The Role of Irrigation Districts in the Transition from Inca Cuzco to Spanish Cuzco. In *Irrigation at High Altitudes: The Social Organization of Water Control Systems in the Andes,* ed. W. P. Mitchell and D. Guillet, 69–98. Society for Latin American Anthropology Publication Series, Vol. 12. Washington, D.C.: American Anthropological Association.

Sheridan, Thomas E.

1988   *Where the Dove Calls. The Political Ecology of a Peasant Corporate Community in Northwestern Mexico.* Tucson: University of Arizona Press.

Sven, Herman

1986   *Tuteños, chacras, alpacas y Macones.* M.A. thesis, University of the Netherlands.

Treacy, John M.

1989   Agricultural Terraces in Peru's Colca Valley: Promises and Problems of an Ancient Technology. In *Fragile Lands of Latin America: Strategies for Sustainable Development,* ed. J. Browder, 209–29. Boulder, Colo.: Westview Press.

1994   *Las chacras de Coporaque: Andenería y riego en el Valle del Colca.* Lima: Instituto de Estudios Peruanos.

Ulloa Mogollón, Juan de

1965 [1586]   Relación de la Provincia de los Collaguas para la discripción de las Indias que su majestad manda hacer. In *Relaciones Geográficas de Indias,* Vol. 1, ed. Marcos Jiménez de la Espada, 326–33. Madrid: Biblioteca de Autores Españoles.

Valderrama, Ricardo, and Carmen Escalante

1988   *Del Tata Mallku a la Mamapacha: Riego, sociedad y ritos en los Andes peruanos.* Lima: DESCO.

Part 2

**Methods and Models for Analyzing
Local Irrigation**

6

Rapid Rural Appraisal of Arid Land Irrigation
A Moroccan Example

**John R. Welch, Jonathan B. Mabry,
and Hsain Ilahiane**

Irrigation is a critical element of international development efforts to ensure global food production, and Rapid Rural Appraisal (RRA) techniques can assist agricultural planners in their assessments of the environmental, technological, and social factors that determine the success of irrigated agroecosystems. Despite the obvious importance of irrigation in arid and drought-prone regions, however, few published works are currently available to guide RRA studies in such contexts. In this chapter we describe the methods we employed in a recent RRA study of smallholder farming in southeastern Morocco, discuss the merits and limitations of these methods, and evaluate RRA applications in the planning, implementation, and evaluation phases of the development process. The criteria used in this evaluation include data relevance, reliability, replicability, and comparative potential.

Nowhere is it more evident that irrigation deserves its high priority in development planning than in arid lands, where evaporation far exceeds precipitation and food production is utterly dependent on irrigation (Chambers 1984; Uphoff 1986). In Morocco, for example, irrigation is the foundation for the domestic food supply as well as the export economy, providing 31 percent of all exports and 90 percent of exported agricultural goods (World Resources Institute 1990). The human suffering and economic hardship brought about by the recent North African drought (1979–1986) made it painfully apparent that Moroccan agricultural development must promote irrigation, and that irrigation projects must be planned with consideration of long-term environmental variability (see Stockton 1988; Stockton and Meko 1990). At the same time, however, the uneven performance of large-scale, technologically advanced irrigation projects has demonstrated that social factors can be at least as important as environment and technology in determining the success of an irrigation system (Coward 1980; Uphoff 1986). Arid land farmers have legitimate concerns about any tampering with their water supplies and are wary of proposals to modify irrigation systems that have fed their families for generations. Moroccan water development thus faces a complex cultural matrix as intractable as the region's challenging environment. For these reasons, planning successful irrigation development demands reliable background information on existing systems, as well as insights into the factors that encourage local participation in the development process.

Rapid Rural Appraisal is one promising way to assist irrigation development in arid lands. RRA is an array of methods for quickly assessing environmental factors as well as human conditions, activities, technologies, and perceptions in order to identify successes, problems, and development needs. This article discusses the RRA data collection methods employed in a study of irrigation agroecology in the Ziz Valley of southeastern Morocco (Mabry, Ilahiane, and Welch 1991). RRA proved an excellent approach to comprehending the region's farming ecology and technology, as well as the organizational factors most relevant to development project planning. Despite these merits, RRA should not be an alternative to the in-depth and ongoing analyses required for successful development project implementation and evaluation; it is most useful in the planning stages.

The History of RRA

Until the 1970s, most studies of developing countries' rural situations were broad statistical analyses based on conventional economic assumptions. These analyses focused on national and international policy concerns, seldom considering the details of local economies or ecologies in project regions, or incorporating systematic consultations with local inhabitants (Dahlberg 1990). The less than optimal performance of many development projects that were based on these general analyses made donor agencies wary of projects designed without detailed information on the region, or input from its occupants (Staatz and Eicher 1990).

As the shortcomings of this approach surfaced, development priorities began to shift from policy concerns and "top-down" program implementation to "farmer-first" development for and by the people (Cernea 1985; Chambers 1988, 1990; Brokensha 1989). Development agency personnel gradually realized the need for multidisciplinary data collection strategies sensitive to local circumstances and the benefits of regarding local populations as more than "project background." Planners began to turn to the people for their perspectives on programs, and to define success in terms of local support for their work. Farming Systems Research (FSR), a holistic approach emphasizing assistance to smallholders, began to attract attention (Baker and Norman 1990). In response to the same concerns, early RRA studies impressed development planners by yielding abundant information from limited time and budgets and by displaying its adaptability to varied local circumstances. RRA began to play an increasingly important role in the development planning process, including irrigation development (Chambers 1985; Conway and McCracken 1990).

The emergence of RRA reflects development programmers' mounting interest in the details of social factors of production (Richards 1985). In particular, the trend involves recognition of the value of cross-cultural comparative perspectives, especially the study of households and other basic resource management units (Chambers 1984, 1985; Rhoades 1985). With donor agencies serving more as facilitators than as implementors, the challenge to irrigation development became how to diagnose the advantages and disadvantages of existing water management institutions, irrigation techniques, and cropping practices, how to identify solutions to specific problems, and how to foster local participation in program planning, implementation, and evaluation.

RRA today is widely applied under various rubrics, but it is far from a coherent or consistent method. It has been suggested that RRA is applicable in urban settings, and can serve the interests of long-term research (Frankenberger and Lichte 1985). We prefer a more restricted definition: Rapid Rural Appraisal is a short-term, semi-structured (but systematic) program of acquisition and analysis of information on rural life carried out by a multidisciplinary team. Common denominators of RRA studies include short time schedules, intentional avoidance of biases that result from disproportionate research attention on males, elites, and easily accessible and organized settings during comfortable seasons, emphasis on conceptual clarity rather than on statistical characterization, and continuous, infield refinement of methods in response to new findings (Frankenberger and Lichte 1985; Rhoades 1985; Beebe 1995). Previous applications of RRA to irrigation development have been limited to the humid tropics and subhumid regions (e.g., Chambers 1985; Harvey, Potten, and Schoppmann 1987). We know of no other broadly applicable discussion of RRA techniques for arid land irrigation systems; until now, little attention has been given to the special problems faced by farmers in regions where water supplies are largely a function of capricious climates.

The Ziz Valley Project

Objectives

Our three-person team's research interests centered on the technological and organizational means that Moroccan farmers use to cope with their drought-prone climate and uncertain water supply. The United States Agency for International Development (USAID) gave us seven weeks in the summer of 1991 to pursue these interests while addressing issues raised by a Project Implementation Document (PID) for redesigning aspects of the Tadla irrigation system in central Morocco (USAID 1991). The seven ambitious and overlapping goals of the Tadla PID further focused our project (see Swearingen 1987 for an overview of Tadla development). These goals were (1) enhanced operational flexibility in water delivery, (2) greater irrigation perimeter productivity, (3) improved economic and environmental sustainability, (4) promotion of private sector participation, (5) more efficient use of available resources, especially water, (6) better soil and water conservation, and (7) expansion of commercially oriented, large-scale crop production.

In formulating a project design that would advance these seven operational, economic, and ecological objectives through a short field project, USAID encouraged a methodological focus for our study: the development and evaluation of broadly applicable data collection techniques. We targeted methodological issues confronting planners of arid land irrigation projects in complex social settings and, for reasons already discussed, settled on RRA as the best approach. To make the study relevant to the Tadla objectives, our data collection emphasized five factors central to all irrigation: (1) water supply, (2) technology and technological change, (3) relations of production, (4) efficiency, and (5) sustainability. Because of the large size and daunting complexity of the Tadla irrigation system, we decided to test our methods in the Ziz Valley, on a set of much smaller systems that we thought we could study more effectively during our limited time.

In addition to developing RRA methods for arid land irrigation systems, the project had two secondary goals that were less closely tied to the

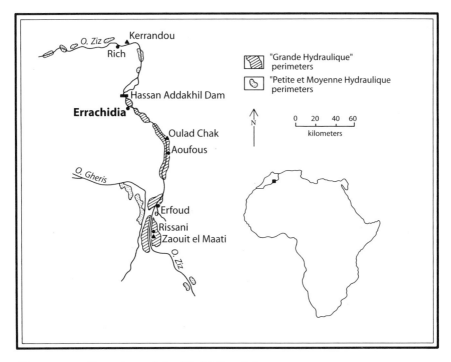

Figure 6.1   Location of the Ziz Valley, Morocco.

Tadla PID. First, we sought to identify Ziz Valley agricultural potentials, constraints, and farmer assistance needs. Second, we examined the introduction of market-oriented crops, technologies, and organizational structures in terms of their probable consequences for the valley's subsistence-oriented agricultural economy.

Project Schedule and Regional Background

Once in Morocco, we spent a week consulting with faculty of the Institut Agronomique et Vétérinaire Hassan II (IAV), finalizing project plans, and conducting background research in the capital, Rabat. From there, we traveled to Errachidia, the largest town in the Ziz Valley (figure 6.1), where we dedicated a second week to compiling secondary data, becoming familiar with the region's irrigation infrastructure, and consulting with the Office Regional de Mise en Valeur Agricole du Tafilalet (ORMVAT). During the subsequent three weeks of work with Ziz Valley farmers, we applied a series of RRA data collection techniques. The remaining two weeks of in-country work entailed preparing a report for USAID-Rabat (Mabry, Ilahiane, and Welch 1991).

The Ziz River Valley is a 250-kilometer-long ribbon of irrigated agricultural land that winds out of the eastern High Atlas Mountains and cuts through the pre-Saharan wastes before being absorbed into the Saharan depression. The Ziz River drains approximately 14,125 square kilometers, a watershed with an average altitude of about 1100 meters. Over 1 million date palms and thousands of olive trees mark the twisting, 0.5- to 10-kilometer-wide valley as it bisects Morocco's province of Errachidia, which encompasses roughly 10 percent of the kingdom. The population of Errachidia practices intensive irrigated agriculture, and both sedentary and seminomadic pastoralism.

The Ziz Valley was an ideal place for our study for both analytical and practical reasons. The technological diversity there is intriguing, with modern and traditional irrigation systems each serving about one-half of the Ziz Valley's agricultural lands (totalling about 50,000 hectares). Traditional systems of small brush-and-stone diversion works and earthen canals have been overlain or supplemented by a large, recently built network of cement dams and canals supplied by a single flood storage reservoir, the Hassan Addakhil. Given our theoretical interests in traditional agroecosystems and agricultural sustainability, we were keenly interested in comparing modern and traditional irrigation infrastructures, cropping patterns, and resource management to better understand both kinds of systems.

In order to ascertain the range of variability in Ziz Valley cropping systems and farmer perspectives, we selected sample communities for intensive data collection in the upper, middle, and lower reaches of the valley (table 6.1; see Welch, chapter 4 this volume, for a similar approach to the analysis of irrigation in the Sefrou region of Morocco). The upper valley community, Kerrandou, lies above the storage reservoir and continues the age-old struggle with the Ziz River's highly variable surface flow. Below the Hassan Addakhil reservoir, Oulad Chak exemplifies villages that now enjoy a relatively stable and moderately saline water supply in the middle stretch of the valley. As the Ziz River continues south onto the Tafilalet Plain, its diminishing flow becomes increasingly polluted and saline. We examined the lower Ziz River community of Zaouit el Maati to understand the problems faced by tailend farmers relying on the river's residual flow.

Table 6.1   Agroecological characteristics of three communities in the Ziz Valley, Morocco.

Community	Kerrandou	Oulad Chak	Zaouit el-Maati
Elevation	1250 m	925 m	750 m
Annual precipitation	200 mm	125 mm	50 mm
Population	1250	1617	765
Water supply	Ziz River	Ziz River, Meski Spring	Ziz River, wells, pump station
Water salinity	low	medium	high
Primary crops	hard and soft wheat, barley, corn, olives, other fruits	soft and hard wheat, barley, dates, olives, other fruits	dates, soft wheat, barley
Major problems	floods	palm fungus	inconsistent water supply, salinization

RRA Data Collection Methods

The data collection techniques we used include methods for obtaining qualitative and quantitative information on cropping systems, farming techniques and technology, and farmer perspectives and opinions. Data collection techniques are seldom equally appropriate from project to project. Some techniques are essential (e.g., interviews); some are applicable wherever sufficiently detailed data can be collected (comparative exercises); other techniques must be carefully developed in response to the requirements of specific projects (questionnaires).

Prior to Fieldwork

Time limits and the large size of the research area mandated that our plans be detailed yet flexible, and that we have data collection instruments ready for use prior to departure for Morocco. Before our departure, we developed an outline of data necessary for basic socioeconomic and ecological

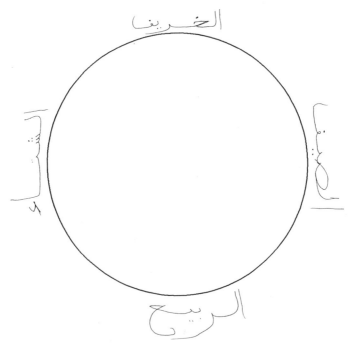

Figure 6.2 Cropping and irrigation calendar diagram used by the Ziz Valley RRA Project.

analyses of irrigation systems, a preliminary interview structure for use in our discussions with farmers, a plot survey form, and, finally, a graphic instrument for diagramming the annual cropping and irrigation calendars (figure 6.2)..In finalizing the project design during our first week in Morocco, we selected the following methods to glean the most information from a short period of fieldwork in diverse situations: background research, key consultant interviews, group interviews, informal consultations, impromptu farmer surveys, plot surveys, photography, crop calendar diagramming, and farmer activity demonstrations. In what follows, we describe each data collection technique and examine the biases that can potentially corrupt their results.

Background research is the accumulation of all manner of data from secondary sources; it helps to focus field investigations on the most appropriate geographical and topical emphases. Our background research took place in the United States and in Rabat, and continued during our fieldwork at ORMVAT headquarters and various ORMVAT Centres de Mise en Valeur (CMVS; agricultural extension offices). We structured our three-community sample and our interviewing schedule on the basis of a three-day reconnaissance of the entire study area and information on farm location, labor demographics, and cropping system variability available from IAV and ORMVAT libraries and archives. These sources also yielded information on long-term trends in precipitation, hydrology, and crop yields.

Field Techniques

Although most of the methods discussed here are widely applied and well understood, the potential for misapplication exists. For this reason, each technique merits a brief description along with an evaluation of its use in our RRA study.

Key consultant interviews are in-depth discussions with individuals possessing information of particular relevance to a project. The individuals we targeted for these interviews were treated as experts. At our request, in addition to substantive data, most of them provided frank and useful assessments of our goals, methods, and preliminary findings.

Group interviews are semistructured conversations with more than one individual. Group interviews proved to be excellent ways to elicit diverse opinions on controversial issues. Although this technique can introduce researchers to new perspectives, it can also lead to tangential discussions. In Morocco, and in other places where farmers are typically eager to share their views, interviews with more than three farmers can be difficult to keep on track. Group interviews are most profitably conducted during early stages of fieldwork, as a means for introducing the project to locals, and for identifying farmers' primary concerns.

Impromptu farmer surveys are nonsystematic efforts to obtain responses to a particular question from a readily accessible sample of individuals. One afternoon, for example, five farmers that we encountered in their fields were polled on the number of irrigation applications required for corn crops. This technique is more appropriate as a means for improving the analytic value of the questions being asked than as a way of obtaining representative responses.

Informal consultations are brief and unplanned conversations with individuals or small groups. Informal consultations with people we met in fields, villages, and other contexts proved a valuable way of identifying issues and data sources that subsequently became central aspects of the project. The unstructured nature of these consultations led to discussions not always subsumed by more formal data collection, alerting us to issues that emerged as particularly important in the course of the study. For example, a walk through fields near one village provided the opportunity to discuss the relative merits of manuring and chemical fertilization with farmers applying the two different nutrients at the same time in adjoining fields. Chemical fertilizers are being "pushed" by many extension agents (in Morocco and elsewhere), and this informal session opened our eyes to the benefits and costs of forsaking the more traditional manure applications.

Plot surveys record data on the physical and biological characteristics of agricultural plots (e.g., plot dimensions, technical features, crops, and economic trees). We used a recording form with questions on one side and a map grid on the back. The grid was used in sketching the plot and its contents. Also, because this technique requires only observational skills, it allows less linguistically proficient team members to contribute significantly to data collection.

Farmer activity demonstrations are responses to requests to "show us how that is done." These demonstrations can clarify points made orally, and can provoke additional questions. They also create opportunities for team members to gain appreciation for the many skills involved in local agriculture.

Photography provides a visual record of the people, places, and features relevant to fieldwork. We employed aerial photographs collected during background research to establish our location on larger-scale topographic maps, and to identify spatial relationships not apparent on the ground. We shot a parallel set of color slides (for presentations and archival purposes) and black-and-white prints (for publications). Color prints were taken when consultants or acquaintances showed an interest in being photographed. Our willingness to snap pictures was seen as a sign of friendship, and typically increased individuals' responsiveness to our inquiries. Our summer 1991 photographic documentation is available for comparison with records from other seasons and years (past and future) with the goal of gaining insight into the dynamics of field, river channel, and social conditions before and after the region's irrigation development (Welch, Mabry, and Ilahiane 1996).

Cropping and irrigation calendar diagramming is the elicitation of information on the typical annual calendar of agricultural activities. We employed a graphic data collection instrument showing the annual cycle of seasons marked on a large circle (figure 6.2), but linear graphs or other formats may be more in tune with conceptions of time elsewhere. Some farmer consultants constructed pie-type charts, while others elected to draw concentric bands to depict the timing of critical irrigation activities. Although we had intended to use this method only with farmers, it was also employed to elicit data on extension agents' annual activity cycle, and it facilitated consultations with engineers about the annual schedule of reservoir water releases. Comparison of the different responses of these three participants in Ziz Valley agriculture (farmers, extension agents, and engineers) highlighted differences in perceptions of seasonal variations in irrigation activities and cropping patterns.

Other diagramming. Whenever a consultant (or team member) was interested in expanding or clarifying a particular point, we provided a piece of paper and a pen to facilitate expression. Responses ranged from simple lists to complicated diagrams. Some farmers drew diagrams on the ground instead.

Potential Sources of Bias in RRA Studies

The following discussion of factors that threaten the integrity of RRA data collection techniques reflects specific situations our team encountered in the study area. Nonetheless, the comments are intended to assist researchers in formulating data collection strategies for other regions.

Institutional Affiliation Biases

Institutional affiliation can become a source of bias when there are overlapping jurisdictions in host countries. National or local administrators often have cultural, political, or administrative motives for seeking to limit the time or territory available to fieldworkers. The astute researcher will avoid political entanglements by remaining independent while, at the same time, developing substantial connections with a wide range of individuals as well as with local and national organizations (e.g., village councils, rural development and agricultural agencies, universities). Although institutional independence is seldom possible for logistical reasons (e.g., transportation, office support), this freedom should be maintained as long as possible.

Spatial Biases

Spatial sampling is probably the most important problem facing the researcher applying RRA methods. Locational factors have pervasive influences on the representativeness and quality of the data collected. To avoid spatial biases, wherever practical, get away from extension offices; go beyond paved roads and settlements with modern amenities; break out of pre-chosen sample areas.[1] In irrigation systems, visit the head, middle, and tail of the cultivated areas.

Temporal Biases

Time-sampling problems are perhaps the most difficult to overcome. Few projects have access to data sets containing similar information for each distinct cropping season, much less for previous years. Because of the importance of these data for most development concerns (e.g., sustainability), it is imperative that researchers seek information on the entire calendar year, last year, the best year, the average year, and the worst year. If possible, visit the system at different seasons of the year, days of the week, and times of the day. Avoid rigorous scheduling unless it is absolutely necessary; wander the system. Investigate discontinuities between reports and observations. Oral history is an excellent source of information on long-term environmental and system changes, provided that the testimony can be cross-checked and anchored to specific dates (e.g., years remembered for severe droughts, floods, and pest infestations, which are often exaggerated in their retelling).

Personal Biases

Sampling problems related to individual roles and perceptions are also difficult to avoid when working in an RRA framework. Although it is important from the standpoint of community relations and breadth of understanding to obtain data from many sources, depth of understanding must be established by targeting primary sources for the data critical to a project's focus. Volunteered testimony should be listened to, but then scrutinized in terms of the motivations of individuals eager to contribute and in terms of the individuals' qualifications to speak as experts on particular topics.

Our approach to Ziz Valley farming explicitly assumed that subsistence farmers, rather than administrators, engineers, or extension agents, were the real experts. Despite impressive knowledge and sincere intentions, extension agents are seldom farmers themselves. They are often unfamiliar with the details of farming systems so critical to long-term agricultural success. We avoided infield participation in our data collection by extension agents, engineers, and bureaucrats. These individuals may seek to promote special interests or programs, and their presence invariably influences farmers' responses. Ziz Valley extension agents, for example, often answered questions we had directed at farmers, arguing that farmers "do not know how to say it right." We were particularly wary of agents' influences when they disparaged indigenous know-how. However, if researchers wish to understand the bureaucratic culture in which extension agents operate, and to identify problems between administrators and farmers, they should also interview the bureaucrats.

Skill Biases

Variability in skills within the research team constitute a significant problem in the typically multidisciplinary RRA framework. Because of the importance of collecting appropriate data on the first attempt, the strengths and weaknesses of team members' data collection skills have to be built into the division of labor. Language skills are particularly critical in rural Morocco. Although a great deal of useful data on material and physical aspects of irrigation farming can and should be collected by team members uninfluenced by oral cues, participation in interviews that demand detailed information about everything from techniques to perceptions of differences between past and present, require that at least one team member be 100 percent fluent in the local dialect(s).

Although this point is obvious to most fieldworkers, it bears repeating that farmers seldom think like interviewers or administrators, engineers, and extension agents. Academic jargon does not translate well. Simple questions presented in the local idiom are essential for collecting clear and relevant data.

RRA Potentials and Limitations

Next we discuss the costs and benefits of an RRA approach to collecting information for arid land irrigation development by examining the results of our work in terms of data relevance, data reliability and replicability, and the comparative potential of the research. These three qualitative dimensions of variability constitute a means for evaluating the success of data collection efforts, and for making observations on the overall utility of RRA as a tool for development planners.

Data Relevance

The relevance of data can be measured by the degree to which the types of information collected meet the analytic needs that prompted the inquiry. In general, because RRA is a broad research strategy that gives researchers great latitude, RRA generally rates well in terms of relevance. Because our objectives were primarily methodological, however, a brief reconsideration of the analytic needs outlined in the Tadla PID (USAID 1991) identifies the categories of data we pursued and reveals that the Ziz Valley project results were relevant in terms of developing methods potentially useful in the Tadla region. In light of the limited amount of social and ecological data available for the Tadla area, the first step might involve defining the range of geographical, technological, and social variability in the region. RRA is ideal for such purposes. A second phase might determine the modal strategies (for water use, cropping, seeking credit, and so on) employed by the major subsets of Tadla's farmers. A third step could then evaluate common strategies in terms of risk minimization, profit maximization, and long-term ecological sustainability in order to identify general problem areas, understand deviations from general patterns, and formulate development programs.

We are confident that the RRA approach is well-suited to contribute to at least the first two steps previously outlined. Our data collection efforts successfully identified the basic range of variability and the distinctive elements of the three Ziz Valley irrigation systems studied. An additional month of fieldwork would have been required for us to rigorously define the modal strategies adopted by groups of farmers from the three regions, but we think that the methods employed could have been easily adapted to serve this goal. In contrast, the third goal of strategy evaluation and program formulation probably would not be best served by an RRA approach. RRA is a poor substitute for the detailed, long-term research necessary for large-scale irrigation development planning. RRA is most appropriate in initial phases of projects.

Data Reliability and Replicability

Often viewed as the hallmarks of science, data reliability and replicability lose much of their analytic value outside of the laboratory. Reliability refers to the precision and accuracy of the information, while replicability is, in the context of development studies, a concept for determining whether questions about whether the data collected represent the spatial, temporal, and behavioral variability within the study universe. Regrettably, some of the same qualities that make RRA most attractive to development planners (flexibility, efficiency, problem-orientation) may limit the reliability and replicability of RRA data. For example, only on rare occasions was our team able to pose follow-up questions to individual farmers or to cross-check information supplied by a particular extension agent or farmer during subsequent consultations. The "hit-and-run" nature of RRA similarly constrains researchers' capacity to ascertain the representativeness of their results for the region as a whole, for other times of the year, and for individuals not contacted. This has become a serious problem in other projects. The solution is clear: Development initiatives must be prepared for long-term commitment to, and involvement with, the communities of people they seek to assist. As an example, Uphoff (1988) details an enlightened program for farmer participation in project design and implementation and, far more importantly, in project evaluation and redesign.

Comparative Potential

Whether the data collected are of use to other projects and are organized in a way that allows comparison with data from other projects determine its comparative potential. Comparability is becoming an increasingly important criterion for evaluating the results of development studies as publication outlets increase in number and scope, as planners realize the benefits of learning from previous mistakes, and as development research becomes more closely articulated with mainstream concerns in various academic disciplines. Although it is probably beneficial for researchers to maintain an eye toward comparability, their primary goal should be to collect data relevant to immediate project objectives. This means that data collection methods successful in one project should not be applied in "cookbook" fashion elsewhere. On the other hand, the collection of fundamental types of empirical data should be a universal aspect of all agricultural development projects. For arid land irrigation studies in particular, all researchers should give high priority to obtaining information on social relations of production, precipitation patterns, the size of field systems, water quality and quantity, the annual cycle of cropping practices, and yields.

For the Ziz project, collecting these fundamental data in the field and comparing them with published data revealed some intriguing parallels between Ziz Valley agriculture and the productivity and sustainability of agricultural systems elsewhere in Morocco, and in other world regions (Mabry, Ilahiane, and Welch 1991; Mabry and Cleveland, chapter 11). For example, we employed data compiled by extension agents as well as farmer interviews to examine agricultural costs and returns in the Ziz Valley. Comparisons with other regions indicated that the Ziz Valley is a reasonable model for what sustainable irrigated agriculture looks like in Morocco, and the model merits emulation as well as further study. The data also demonstrate that Ziz Valley farmers make many decisions on the basis of nonmarket objectives. For example, 60 to 100 percent of wheat production along the Ziz River is primarily for household consumption. Labor resources usually come from within the family or arranged labor groups (see Netting 1993). When outside labor is necessary, payment is often a proportion of the harvest, rather than cash. These patterns strongly suggest that Ziz Valley farmers seek, in general, to minimize risks rather than maximize profits. In the drought-prone pre-Sahara, stable yields and agricultural sustainability are more important than short-term productivity.

RRA proved extremely valuable in identifying these patterns, and is ideally suited for the collection of the parallel data sets required to expand these sorts of comparative analyses. The efficiency and resilience of the Ziz Valley's farmers indicate that new programs for improving resource management and encouraging more effective credit distribution — at Tadla and elsewhere — should encourage and incorporate community input, building on the successes of traditional patterns while making innovative strategies available to farmers through pilot and demonstration programs (Moles 1989).

The Potentials of RRA

What RRA fails to provide in terms of reliability and replicability it can make up for in flexibility, relevance, and data comparability. The RRA strategy provides a wealth of data collection options, many of which are useful for arid land irrigation development studies. Linkages between farmers' problems and regional development objectives are vital to project success, and RRA is well-equipped to assist in defining these links, and to determine the appropriateness of development project goals and methods. Such research on smallholders and existing irrigation institutions can provide models for efficient and sustainable resource management.

Although RRA studies that tap into local funds of knowledge are a step toward the goal of grassroots development, achievement of full farmer participation requires that planning personnel signal long-term commitment to local communities, rather than to the project, per se. The local level of interest in development is likely to be inextricable from their broader concerns with controlling critical resources, minimizing risks, and influencing project decision making to meet their changing needs. The most effective means for determining whether a particular innovation holds promise for an alternative context is to introduce it to farmers, and let them decide whether or not to use it (Richards 1985). This strategy demands that the unfortunate communication gap that separates the scientific community and rural farmers be overcome, and that scientists take the lead in this regard. Traditions should be treated as reflections of tried-and-true strategies rather than as impediments to modernization and progress (see Abdellaoui 1987; Chambers 1988, 1990). As a flexible and effective approach to getting to know the people, places, and customary farming of a region targeted for a development initiative, RRA is an appropriate means for bridging the gap between innovations and more traditional farming techniques and community concerns.

Acknowledgments

The research reported here was a joint study by the University of Arizona Bureau of Applied Research in Anthropology (BARA) and the Institut Agronomique et Vétérinaire Hassan II (IAV), sponsored by USAID-Morocco. The project was made possible by our associates at the IAV and the Tafilalet Office Regional de Mise en Valeur Agricole (ORMVAT). The IAV director, M'hamed Sedrati; the ORMVAT director, Mohammed Hajjaji; the governor of

Errachidia, the Honorable Ahmed Arafat; as well as Abdelhak Guemimi, Charles Stockton, Abdellah Hammoudi, and Thomas Park, were all essential to the project's success. Special thanks to Moha Boumezough. Carlos Vélez-Ibañez, former director of BARA, provided essential administrative support. Comments made on an earlier draft of the article by Robert Netting, Timothy Finan, and Thomas McGuire proved especially useful. Responsibility for errors resides solely with the authors.

Note

1. For safety reasons, and to avoid "diplomatic incidents," researchers should pay close attention to the dynamics between government officials and farmers and to local politics before exploring unselected areas or getting away from government escorts.

References Cited

Abdellaoui, R. M.
   1987   Small-Scale Irrigation Systems in Morocco: Present Status and Some Research Issues. In *Public Intervention in Farmer-Managed Irrigation Systems,* 165–73. Digana, Sri Lanka: International Irrigation Management Institute.
Baker, Doyle C., and David W. Norman
   1990   The Farming Systems Research and Extension Approach to Small Farmer Development. In *Agroecology and Small Farm Development,* ed. M. A. Altieri and S. B. Hecht, 91–104. Boca Raton: CRC Press.
Beebe, James
   1995   Basic Concepts and Techniques of Rapid Appraisal. *Human Organization* 54(1): 42–51.
Brokensha, David
   1989   Local Management Systems and Sustainability. In *Food and Farm: Current Debates and Policies,* ed. C. Gladwin and K. Truman, 179–97. New York: University Press of America.
Cernea, Michael
   1985   *Putting People First: Sociological Variables in Rural Development.* Oxford University Press, Oxford.
Chambers, Robert J. H.
   1984   Beyond the Green Revolution: A Selective Essay. In *Understanding Green Revolutions: Agrarian Change and Development Planning in South Asia,* ed. T. P. Bayliss-Smith and S. Wanmali, 362–79. Cambridge: Cambridge University Press.
   1985   Shortcut Methods of Gathering Social Information for Rural Development Projects. In *Putting People First: Sociology and Development Projects,* ed. M. Cernea, 399–415. Washington, D.C.: World Bank.
   1988   *Managing Canal Irrigation: Practical Analysis from South Asia.* Cambridge: Cambridge University Press.
   1990   Farmer-First: A Practical Paradigm for the Third Agriculture. In *Agroecology*

and Small Farm Development, ed. M. A. Altieri and S. B. Hecht, 237–44. Boca Raton: CRC Press.

Conway, Gordon R., and Jennifer A. McCracken

    1990   Rapid Rural Appraisal and Agroecosystem Analysis. In *Agroecology and Small Farm Development,* ed. M. A. Altieri and S. B. Hecht, 221–35. Boca Raton: CRC Press.

Coward, E. Walter, Jr. (editor)

    1980   *Irrigation and Agricultural Development in Asia: Perspectives from the Social Sciences.* Ithaca: Cornell University Press.

Dahlberg, Kenneth

    1990   The Industrial Model and Its Impacts on Small Farmers: The Green Revolution as a Case. In *Agroecology and Small Farm Development,* eds. M. A. Altieri and S. B. Hecht, 83–90. Boca Raton: CRC Press.

Frankenberger, Timothy R., and John L. Lichte

    1985   *A Methodology for Conducting Reconnaissance Surveys in Africa.* Farming Systems Support Project, Networking Paper No. 10. OALS.

Harvey, J., D. H. Potten, and B. Schoppmann

    1987   Rapid Rural Appraisal of Small Irrigation Schemes in Zimbabwe. *Agricultural Administration and Extension* 27:141–55.

Ilahiane, Hsain

    1991   Small-Scale Irrigation and Social Inequality in the Ziz Valley, Southeast Morocco. Unpublished manuscript in author's possession.

Mabry, Jonathan B., Hsain Ilahiane, and John R. Welch

    1991   *Rapid Rural Appraisal of Moroccan Irrigation Systems: Methodological Lessons from the Pre-Sahara.* Report submitted to United States Agency for International Development, Rabat, Morocco.

Moles, Jerry A.

    1989   Agricultural Sustainability and Traditional Agriculture: Learning from the Past and Its Relevance to Sri Lanka. *Human Organization* 48(1):70–90.

Netting, Robert McC.

    1993   *Smallholders, Householders: Farm Families and the Ecology of Intensive, Sustainable Agriculture.* Stanford: Stanford University Press.

Postel, Sandra

    1989   *Water for Agriculture: Facing the Limits.* Paper No. 93. Washington, D.C.: Worldwatch Institute.

Rhoades, Robert E.

    1985   Informal Survey Methods for Farming Systems Research. *Human Organization* 44(3):215–18.

Richards, Paul

    1985   *Indigenous Agricultural Revolution: Ecology and Food Production in West Africa.* London: Hutchinson.

Staatz, John M., and Carl K. Eicher

    1990   Agricultural Development Ideas in Historical Perspective. In *Agricultural De-*

velopment in the Third World, ed. C. K. Eicher and J. M. Staatz, 3–38. Baltimore: Johns Hopkins University Press.

Stockton, Charles W. (ed.)

1988   *Drought, Water Management and Food Production.* Mohammedia, Morocco: Imprimerie Fedala.

Stockton, Charles W., and David M. Meko

1990   Some Aspects of the Hydroclimatology of Arid and Semiarid Lands. In *Human Intervention in the Climatology of Arid Lands,* ed. D. R. Haragan, 1–26. Albuquerque: University of New Mexico Press.

Swearingen, Will D.

1987   *Moroccan Mirages: Agrarian Dreams and Deceptions, 1912–1986.* Princeton: Princeton University Press.

United States Agency for International Development (USAID)

1991   *Water and Soil Resources Development (Project #608-0213).* Project Identification Document Prepared by USAID-Morocco, Office of Agriculture and Natural Resources, Rabat, Morocco.

Uphoff, Norman Thomas

1986   *Improving International Irrigation Management With Farmer Participation: Getting the Process Right.* Boulder, Colo.: Westview Press.

1988   Participatory Evaluation of Farmer Organizations' Capacity for Development Tasks. *Agricultural Administration and Extension* 30:43–64.

Welch, John R., Jonathan B. Mabry, and Hsain Ilahiane

1996   *The Agroecology of a Moroccan Oasis.* Slide set and guide accepted by Pictures of Record. In preparation.

World Resources Institute

1990   *World Resources 1990–91.* New York: Oxford University Press.

7

Simulation Modeling of Balinese Irrigation

J. Stephen Lansing

For over a thousand years, Balinese rice farmers have engaged in co-operative agricultural practices. This remarkable achievement in sustainable agriculture is surprising given the apparent absence of any centralized coercive control mechanism, and water supply conditions that could easily lead to a rapid breakdown of cooperation. For the past decade I have collaborated with a systems ecologist, James N. Kremer, in an analysis of the role of the traditional system of "water temples" in the management of water in Balinese rice terraces. Previously, conversations with Balinese farmers and my own observations had persuaded me that the regional coordination of cropping patterns by water temples had significant effects on rice production. In 1987, I began to work with Dr. Kremer to develop computer simulation models of Balinese irrigation systems to evaluate the productive role of water temples.

We have reported the results of this research in several publications.[1] The purpose of this chapter is not to comment further on these results, but rather to clarify the methodology. My aim is to demonstrate the potential usefulness of simulation modeling as a tool for social scientists interested in the relationship between social and ecological factors in traditional irrigation systems. For this reason I will not discuss the predictive model of two rivers and 172 local irrigation systems that was used to analyze the role of water temples in a particular region of Bali. Instead, I will consider the workings of a simple three-field model based on the same logic as the 172-field empirical model. The two models differ only in that the 172-field model includes empirical values for all variables, whereas the simple three-field model is a conceptual model with the simplest possible values for all variables.

The model described here was built using a modeling program called Stella™.[2] A complete description of the Stella™ model is included at the end of this paper, so that anyone with access to the Stella™ software can reproduce the model for themselves, or construct a similar model in some other computer language. It is, however, not necessary to recreate the model to make sense of the paper.

What kinds of insights can be gained by creating simulation models of traditional irrigation systems? Irrigation systems may develop over a period of decades or centuries in response to shifting patterns of social and ecological influences. In any given case, it may not be possible to discover how much of the resulting pattern of organization is the result of deliberate planning, and how much is a trial-and-error response to local conditions. Even securing a rigorous understanding of the present-day dynamics of water management in a given case may appear to be out of reach because of the sheer complexity of the problem. Hydrological models of the type constructed by engineers and hydrologists require precise data on such topics as rainfall, subsurface geology, evapotranspiration, groundwater flows and catchment areas.[3] Similarly, crop models (the responsiveness of grain crops to conditions such as nutrient flows) and pest models (the population dynamics or growth characteristics of pests) are usually[4] formulated by agronomists as detailed mechanistic models.[4] And social scientists know that the social relationships involved in traditional irrigation systems are often very complicated. For these reasons, it might appear that a deterministic model that could predict, for example, differences in harvest yields based on different management scenarios would be immensely complicated, assuming that it could be constructed.

Systems ecological simulations differ from the models used by hydrologists or plant biologists, however, in that these models are used only to simulate the effects of detailed mechanistic processes. For example, I ignore most of the specific biology of rice pests and model only two factors: their speed of diffusion (the rate at which they move across the landscape), and the amount of damage they cause to rice plants. This simplification is not only permissible, but appropriate if our interest lies not in the differences between, for example, the effects of various types of insects and virus pests on different rice varieties, but in the aggregate effect of pest populations on rice yields at the watershed scale.

What can we learn from simulation modeling at the watershed scale? This is the appropriate level of analysis if our interest is not in plant biology or tertiary irrigation flows, but rather in the effects of the social management of cropping cycles in a watershed that includes more than one or two irrigation systems. The purpose of creating such a model is to try to capture the underlying dynamics of the system at this scale, and by so doing gain insight into the relationship between social systems of control and their ecological effects. Once a model has been constructed, it can be tested in two ways: model predictions can be compared with real data, and the dynamics of the model itself can be analyzed through a process called sensitivity analysis. Experiments that would be impossible to perform in the real world can be conducted in the model world, to help the modeler evaluate how well her conceptual model of the system explains the range of observed behavior in the real world.

Traditional Irrigation and
the Green Revolution in Bali

Our interest in trying to find a way to analyze the ecological effects of the traditional management system was stimulated by the crisis in rice production triggered by the advent of the "Green Revolution" in Bali. During the 1950s, Indonesia (figure 7.1) was forced to import nearly a million tons of rice each year. The government of Indonesia was thus very receptive to the promise of higher yields from the "Green Revolution" in rice, and in 1967 launched a major program called BIMAS (Bimbingan massal, or "massive guidance") to boost rice production by providing farmers with high-yielding rice seeds and access to fertilizers and pesticides. The new rice varieties grew faster than the native crops, and farmers were encouraged to triple-crop the new rice whenever possible. BIMAS reached Bali in 1971, and by 1977 about 70 percent of the rice terraces in South Bali were planted with Green Revolution rice. At about the same time, the Asian Development Bank began a major irrigation development project in Bali. Rice production increased, but as early as 1974 field-level agricultural officials in Bali were reporting "chaos in water scheduling" and "explosions of rice pests."[5] In 1984 I wrote an unsolicited report to the Asian Development Bank, in which I tried to show how these problems were linked to their disruption of the traditional system of water management.

My report to the bank emphasized the ecological role of water scheduling and pest management by water temples. Before the Green Revolution, Balinese farmers met annually in regional water temples to set cropping patterns, which often involved staggering irrigation schedules from one irrigation system to the next. Ritual ties between water temples emphasized the interdependency of upstream-downstream relationships, and the temples also helped solve quarrels over water rights. With the ar-

Figure 7.1   Location of Bali, Indonesia

rival of the Green Revolution, religious ceremonies continued to be held in the water temples, but farmers were encouraged to plant rice as often as possible and the temples lost control of cropping schedules. Yet these traditional schedules had important effects on both water sharing and pest control. By enabling the farmers to synchronize cropping patterns, the temple networks provided a mechanism to facilitate water sharing, and also enabled the farmers to synchronize harvests and thus create brief fallow periods over large areas, thus reducing rice pest populations by depriving pest populations of their habitat. The success of fallow periods as a pest control technique depended on the extent and duration of the fallow period. Unless all of the fields in a large area were fallow at the same time, pests could simply move from field to field. I urged the Asian Development Bank to consider trying to strengthen (rather than weaken) this traditional system of water temple scheduling as an ecologically sound system of water management. But my advice was formally rejected by the director of irrigation and rural development projects in a memorandum to the vice-president of the bank in 1984:

> We do not fully share the expressed concerns of Mr. Lansing. Certainly there is a direct relationship between large areas of fallow land for a considerable period and the population of pests. However, pest control programs carried out efficiently and effectively will control the pest population and allow growing of rice year-round if adequate water resources are available as is done, for example, in certain areas of Central and East Java where farmers grow three rice crops per annum. It should be noted, that there is no development without affecting traditional systems or customs. Everybody can criticize and damage a project, but only a few people can overcome those difficult problems and make the project viable.[6]

The Asian Development Bank continued to advocate the use of pesticides rather than synchronized fallow periods as the right way to control pests. Four years later, a study by World Bank officials reported that the use of pesticides had by then "pervasively polluted the island's soil and water resources."[7] Meanwhile, I had become convinced that the water temple system had evolved to optimize the tradeoffs between water sharing and pest control in different regions of Bali. In the traditional farming system, groups of *subaks* (local irrigator associations) belonging to local water temples adjust cropping patterns cooperatively to achieve fallow periods over sufficiently large areas to minimize dispersal of pests, but coordination of rice planting over too large a scale would create inefficient peaks of water demand. Simulation modeling provided a way to analyze this trade-off, and in so doing gain a deeper insight into the effectiveness of the temple system as a system of ecological management.

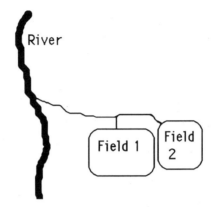

Figure 7.2 A model of two adjacent fields.

Modeling: First Steps

The modeling process begins with an attempt to represent the processes that determine the behavior of the system. We begin with a model of a weir, or diversionary dam, that leads part or all of a river flow into a canal leading to two adjacent field (figure 7.2).

The amount of water in the river depends on rainfall and groundwater flow. In our model, Q1 represents the amount of rainfall that goes directly into the river as runoff from the catchment area. Q2 is the flow into the irrigation canal; Q3 is groundwater flow, and Q4 is the proportion of the flow that goes into other irrigation systems on the other side of the river (figure 7.3).

The amount of rainfall in Bali varies seasonally, which has important implications for irrigation in the dry season. We can simulate this seasonal effect with a simple equation for annual rainfall (see appendix for all equations). The result is a simple hydrological model where daily flow into our irrigation system (Q2) is the sum of rainfall-runoff and groundwater flow (figure 7.4).

The next step is to simulate the growth of rice in the fields. For the sake of simplicity, we assume that rice will grow in the fields until it is harvested, as long as there is sufficient water. Insufficient water for irrigation will cause a proportional reduction in rice growth. In figure 7.5, note that each harvest does not remove all of the rice. In the real world, after harvest much of the plant is left in the fields, where it continues to provide a

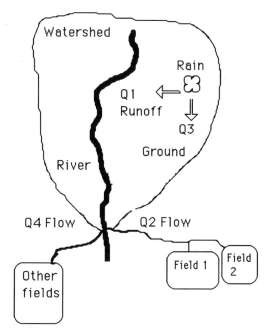

Figure 7.3 Hydrology model.

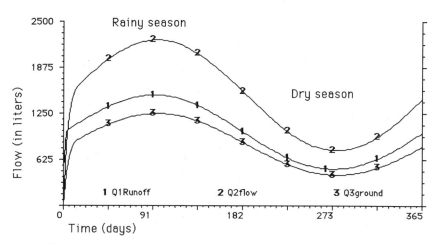

Figure 7.4 Annual rainfall, irrigation flow, and groundwater flow.

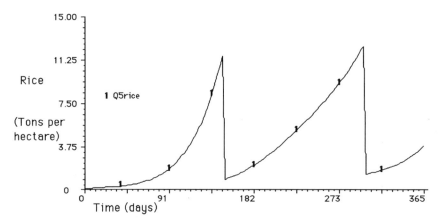

Figure 7.5   Rice growth.

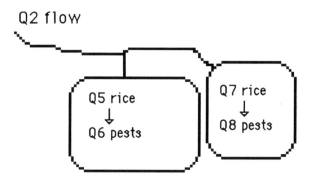

Figure 7.6   Water makes the rice grow, pests eat the rice.

habitat for rice pests such as insects or diseases. Because we are interested in pests, we can simulate this effect in the model by harvesting only 95 percent of the rice.

To model pests, we postulate that there is a small background level of pest infestation on the first day of the simulation. Subsequently, the pest population will grow as long as there is rice for them to consume. Harvests reduce pest populations by removing most of the rice from the fields (figure 7.6).

In the model, the pest population gradually declines in the first six weeks of the simulation because of the lack of food (rice). Thereafter, the pest population tracks the growth of the rice crops. The second harvest is smaller than the first because of the larger pest population (figure 7.7).

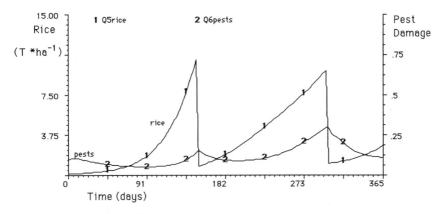

Figure 7.7 Effects of pests on rice harvests.

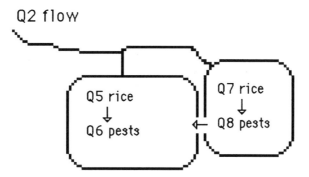

Figure 7.8 Pests diffuse from Field 2 to Field 1.

But pests are not only growing in field 1; they are also growing in field 2, and will spread from one field to its neighbor (figure 7.8).

In figure 7.9, I calculate the effects of the diffusion of pests from field 2 to field 1. Note that the harvest of rice (Q7) for field 2 is higher than the rice harvest for its neighbor (Q5), because some of the Q8 pests are eating Q5 rice. As the model run continues, this effect is enhanced for the second harvest, because the pest population and pest diffusion continue to increase.

Having developed a method to model pest damage and pest diffusion, attention turns to the question of water sharing. If the flow downstream to the Q4 paddies is increased by a small amount, from 25 to 30 percent of the runoff, the harvests of Q5 and Q6 are substantially reduced. The first harvest of Q5, for example, declines from 11 to about 7.4 tons per hectare (figure 7.10).

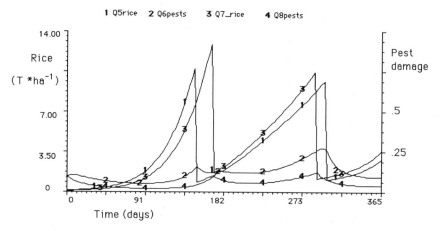

Figure 7.9 Effects of pest diffusion into Field 1 on rice yields.

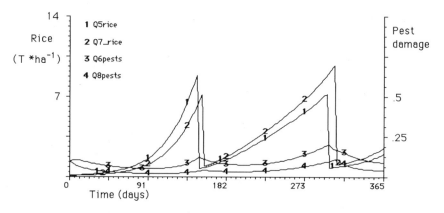

Figure 7.10 Reduction in yields caused by diverting 5 percent additional runoff to downstream neighbors.

This result indicates that the model is very sensitive to differences in water sharing. To make the point even clearer, suppose we increase the share of the runoff to the Q4 paddies downstream from 30 to 50 percent of the flow. The result is a very drastic reduction in Q5 through Q8, as shown in figure 7.11.

By staggering planting dates so as to share available water with downstream neighbors (Q4), however, water stress for Q5 and Q6 can be eliminated. As long as the share of water released to the downstream fields does not exceed 25 percent of the runoff, there is enough water for all. Using

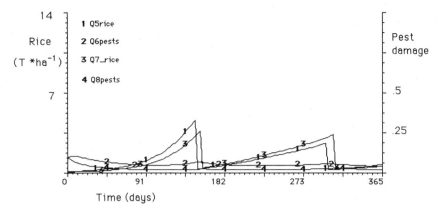

Figure 7.11   Effects of doubling the share of irrigation runoff to downstream neighbors.

the equations for the model that are listed in the appendix, the differences between the graphs in figures 7.9, 7.10, and 7.11 are solely due to changing the J4 equation from J4 = .25 × Q1Runoff (figure 7.9) to J4 = .3*Q1Runoff (figure 7.10) and J4 = .5*Q1Runoff (figure 7.11).

Similarly, pest damage can be minimized by setting a cropping pattern that will produce a fallow period extending over all of the adjacent fields. The optimal solution for this trade-off between water sharing and pest control depends on local conditions, and will not be identical for all regions. If there is plenty of water at all times, a uniform cropping pattern will produce the highest rice yields by minimizing pest damage. But if water is limiting, some offsetting of planting dates may produce the best yields.

This simple model yields a basic insight into the ecological role of the water temple system. If we paid no attention to the relationship between fallow periods and pest populations, it would seem that the upstream subaks would lack any incentive to cooperate with their downstream neighbors. But if we take the role of pests into account, we can see that upstream subaks have much to gain by cooperation. Helping their downstream neighbors satisfy their water needs is likely to be in their own self-interest, because it will allow them to synchronize irrigation schedules and reduce losses from pests. The model has the surprising result that under certain conditions, increasing the pests can lead to higher average rice yields. This can be seen by formulating the trade-off between water sharing and pest control as a Prisoner's Dilemma.[8]

Table 7.1 Prisoner's dilemma payoff matrix for subak cooperation.

	Cooperate (synchronize)	Defect (stagger plantings)
Upstream subak	1	$1 - \pi$
Downstream subak	$1 - \delta$	$1 - \pi$

π represents damage caused by pests
δ represents damage caused by water shortage

Assume that there are two subaks, one upstream and one downstream. The water supply is adequate for both subaks if they stagger their cropping pattern. But if both subaks plant at the same time, the downstream farmers will experience water stress and their harvests will be somewhat reduced. Assume further that pest damage will be higher if plantings are staggered (because the pests can migrate from one field to the next), and lower if plantings are synchronized. This situation is modeled in table 7.1 as a simple two-person Prisoner's Dilemma payoff matrix.

The upstream farmers do not care about water stress, but their downstream neighbors do (this is the famous "tailender" problem: the farmers at the tail end of an irrigation system are at the mercy of their neighbors upstream, who control the irrigation flow). But the upstream farmers do care about pest damage, and pests, unlike water, are quite capable of moving "upstream." So cooperation will always achieve higher yields for the upstream farmers. Downstream farmers will obviously achieve higher yields by cooperating when $\pi > \delta$. Interestingly, there exists a range of pest levels such that increasing the amount of pest damage in the model can result in higher aggregate yields. When $0 < \delta < \pi / 2$, increasing π will increase the incentive for the downstream farmers to cooperate, and lead to a higher aggregate harvest from the two subaks.

Testing Model Predictions:
The Manuaba Subaks

The simulation model allows us to capture the dynamics of a system in which rice yields in a given field depend upon the trade-off between water sharing and pest control through coordinated fallow periods. But how well does this model correspond to the real world? Figure 7.12 shows the approximate location of two small irrigation systems consisting of eleven

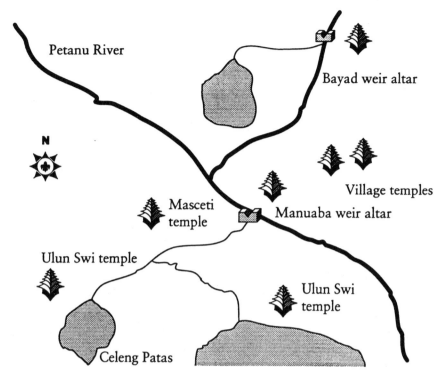

Figure 7.12 Diagram of the Bayad and Manuaba irrigation systems in Southern Bali.

subaks in the upper reaches of the Petanu River in south-central Bali. As the diagram shows, the Bayad weir provides water for a hundred hectares of rice terraces organized as a single subak. Technically, a subak consists of all of the farmers who obtain water from a common tertiary source. A few kilometers downstream from the Bayad weir, the Manuaba weir provides water for about 350 hectares of terraces, organized into ten subaks. Note that in the Manuaba system two subaks, Celeng and Patas, obtain water from one canal, while the remaining eight subaks obtain water from the second canal (table 7.2).

Empirical data on rainfall, irrigation flow, cropping patterns, harvest yields, and pest infestations in this area for 1988 and 1989 were collected by a team of Balinese undergraduate students for the modeling project.[9] All ten subaks in the Manuaba system planted IR 64, a high-yielding Green Revolution rice, in mid-September 1988 and harvested an average of 6.5 tons per hectare in mid-December. Subsequently, they planted

Table 7.2 Sizes of subaks in the Manuaba irrigation system.

Subak	Area (hectares)	Number of members
Gunung	36	57
Patas	42	68
Celang	46	73
Umalawas	22	40
Umalawas Kaja	35	75
Pinyul	35	52
Dukuh	15	32
Munggu	15	40
Penganyangan	35	45
Kendran	69	115

kruing (another high-yielding rice) in early February and harvested 6 tons in May. In June, they all planted vegetables and harvested approximately 2 tons per hectare in August.

The cropping pattern synchronized harvests for all ten subaks, encompassing 350 hectares of rice terraces, thereby possibly helping to keep down pest populations. Pest infestations for this period were reported to be minimal: less than one percent damage to the rice crops, primarily from the brown planthopper. This compares to pest losses of up to 50 percent of the crop in the late 1970s, when subaks planted rice continuously and cropping patterns were very disorganized.

The average flow of approximately 3 liters per second per hectare is low, and suggests that the Manuaba subaks may have experienced some water stress (figure 7.13). Certainly, there was never any excess water. In that light, it is interesting to note that the Bayad subak upstream followed exactly the same cropping pattern as their downstream neighbors, except that they began two weeks earlier. In general, irrigation demand is highest at the beginning of a new planting cycle, because the dry fields must become saturated. By starting two weeks after their upstream neighbors, the Manuaba subaks could avoid water shortages at the time when irrigation demand peaks. Interestingly, the Bayad subak reported an average yield about one ton per hectare higher than the Manuaba subaks. Unlike Manuaba, Bayad also apparently has more than enough water to meet its maximal irrigation demand.

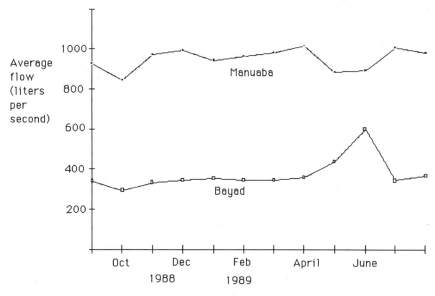

Figure 7.13   Irrigation intakes at the Bayad and Manuaba weirs.

It would appear, then, that the recent history of the Bayad and Manuaba irrigation systems conforms quite well to the assumptions and predictions of our model. By synchronizing their planting schedules, the two groups of farmers minimize pest damage, possibly at the cost of some water shortages at the downstream system. When government agricultural policy forced the downstream farmers to defect, pest losses quickly mounted and provided a stimulus for the farmers to return to the pattern of cooperative synchronized planting schedules. The two-week offset appears to be a bit of clever fine-tuning of irrigation schedules by the farmers, allowing them to offset peak irrigation demand without giving up the synchronized fallow period. (As I have noted elsewhere, the interdependency of the two irrigation systems is given symbolic expression by ceremonial ties between the deities of the two weirs.)[10]

Conclusions

A detailed empirical model that embodies the logic we have developed for the Stella™ model was tested against observed data for 85 percent of the 172 subaks that obtain their water from the Petanu and Oos Rivers.[11] The results showed that most of the observed variation in yields was in

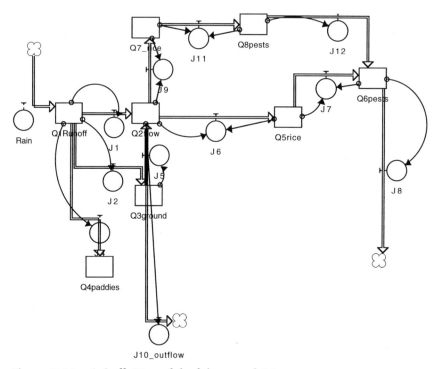

Figure 7.14 A Stella™ model of the two-field system.

close agreement with the predictions of the model. Evidently, the effects of regional coordination of cropping patterns by water temples is more important in explaining variations in yields than more detailed local mechanisms or effects. For this reason, we can say with some confidence that this relatively simple conceptual model (figure 7.14) apparently captures the basic dynamics of the system at the regional scale. Perhaps other researchers may be encouraged by these results to look for other cases in which the collective management of irrigation has important ecological effects.

Notes

1. J. Stephen Lansing, *Priests and Programmers: Technologies of Power in the Engineered Landscape of Bali*, Princeton: Princeton University Press, 1991; J. Stephen Lansing and James N. Kremer, "Emergent Properties of Balinese Water Temple Networks: Coadaptation on a Rugged Fitness Landscape," *American Anthropologist* March 1993; J. Stephen Lansing and James N. Kremer, "Engineered Landscape: Balinese Water Temples and the Ecology of Rice," *Encyclopedia Brittanica Yearbook of Science and the Future*, 1996, Chicago: Encyclopedia Brittanica, 1995.

2. "Stella™ II" is a very easy-to-use computer application for modeling with differential equations, available from High Performance Systems, 45 Lyme Road, Hanover, N.H. 03755; telephone (603) 643-9636. I have often used it to teach ecological modeling to students.

3. Such as the U.S. Army Corps of Engineers HEC series of finite element modeling of shallow water flow, or the Brigham Young School of Engineering FastTABS hydrodynamic models, or the U.S. Geological Survey (USGS) three-dimensional groundwater flow model, MODFLOW.

4. Such as the United Nations Food and Agriculture Organization (FAO) CROPWAT models.

5. J. Stephen Lansing, *Priests and Programmers: Technologies of Power in the Engineered Landscape of Bali*, Princeton: Princeton University Press, 1991, pp. 111–17.

6. Memorandum from K. Takase, Director of Irrigation and Rural Development Department, to Vice President (Projects), Asian Development Bank, 2 October 1984. Provided to me by the Acting U.S. Executive Director of the Asian Development Bank.

7. Badruddin Machbub, H. F. Ludwig, and D. Gunaratnam, "Environmental Impact from Agrochemicals in Bali (Indonesia)," *Environmental Monitoring and Assessment* 11:1–23, 1988.

8. I am indebted to economist John Miller for suggesting the use of the Prisoner's Dilemma to analyze the conditions under which the inclusion of pests in the model may lead to cooperation. Descriptions of the role of the Prisoner's Dilemma in game theory can be found in R. Axelrod, *The Evolution of Cooperation*, New York: Basic Books, 1984; and in Karl Sigmund's excellent popular account, *Games of Life: Explorations in Ecology, Evolution, and Behavior*, Oxford: Oxford University Press, 1993.

9. The students were I Gde Suarja, Ni Made Tutik Andhayani, Dewa Gde Adi Parwata, and I Made Cakranegara.

10. J. Stephen Lansing, *Priests and Programmers: Technologies of Power in the Engineered Landscape of Bali*, Princeton: Princeton University Press, 1991.

11. J. Stephen Lansing and James N. Kremer, "Emergent Properties of Balinese Water Temple Networks: Coadaptation on a Rugged Fitness Landscape," *American Anthropologist* March 1993.

Appendix: Equations for the Stella™ Model of Fields

$$Q1Runoff = Q1Runoff + dt \times (-J1 - J2 - J4 + Rain)$$

$$INIT(Q1Runoff) = 500$$

$$Q2flow = Q2flow + dt \times (J1 + J5 - J6 - J9 - J10 \ \ outflow)$$

$$INIT(Q2flow) = 0$$

$$Q3ground = Q3ground + dt \times (J2 - J5)$$

INIT(Q3ground) = 0

Q4paddies = Q4paddies + dt × (J4)

INIT(Q4paddies) = .03 × Q3ground

Q5rice = Q5rice + dt × (J6 − J7)

INIT(Q5rice) = .1

Q6pests = Q6pests + dt × (J7 − J8 + J12)

INIT(Q6pests) = .1

Q7 rice = Q7 rice + dt × (J9 − J11)

INIT(Q7 rice) = .1

Q8pests = Q8pests + dt × (J11 − J12)

INIT(Q8pests) = .1

J1 = .5 × Q1Runoff

J10 outflow = Q2flow × .5

J11 = .007 × Q7 rice × Q8pests

J12 = .5 × Q8pests × Q8pests

J2 = Q1Runoff × .25

J4 = .25 × Q1Runoff

J5 = Q3ground × .3

J6 = Q2flow × Q5rice × 1e-5

J7 = .007 × Q5rice × Q6pests

J8 = .3 × (Q6pests × Q6pests)

J9 = Q2flow × Q7 rice × 1e-5

Rain = 1000+500 × SIN(2 × pi × time/365).

8

Institutional Innovation in
Small-Scale Irrigation Networks
A Cape Verdean Case

Mark W. Langworthy and Timothy J. Finan

Irrigation networks with multiple users require some sort of institutional framework to address a wide range of management decisions. In particular, decisions about the quantity and timing of water deliveries to each member, procedures for paying for the operation and maintenance of the network, mechanisms to adjust schedules in the face of unforeseen interruptions of operation, and procedures to monitor and punish infractions against operating rules all must be resolved in some manner—either by a coalition of network members, or by an outside agency with authority over the network. Throughout the world a variety of institutional structures handle these management tasks. These structures range from very large, multi-level bureaucracies, such as those in the western United States, to small informal organizations of a few neighboring farmers. These institutions may be distinguished on the basis of the degree to which management functions are undertaken by network members or external agencies. At one extreme are indigenous or local management institutions, in which the investment, management, and regulation activities are all done directly by some coalition of network members. At the other extreme are networks in which all these activities are performed by external (usually national government) agencies whose constituencies transcend the membership of individual irrigation networks. Many irrigation networks around the world are characterized by some combination of local and external control.

The structure and operation of the management institutions have significant impacts on the production decisions and resulting incomes of network members. For example, limitations on access to water imposed by the way in which the network is operated may reduce individual farmers' yields, or even preclude production of crops with more stringent water demands. Analysis of the economic performance (total revenues minus all operating costs) of farmers under existing and alternative irrigation management regimes provides insights into the extent to which the prevailing institutions promote or prohibit the economic well-being of the network members. With this information, two specific questions can be addressed. First, to what extent are economic efficiency goals attained and are the available water resources being used to generate the greatest possible returns for the network as a whole? Second, how do the institutional arrangements divide the returns among the individual members of the network?

In recent years, a growing number of economists have begun to examine how particular institutional structures affect the economic performance of individuals and groups (Davis and North 1971; Hayami and Ruttan 1985). Hayami and Ruttan have elaborated a theory that links the structure of institutions with the pattern of economic and technological change. They argue that "changes in factor endowments, technical change, and growth in product demand have induced change in property rights and contractual arrangements in order to promote more efficient resource allocation through the market" (Hayami and Ruttan 1985:97). In other words, market pressures reflecting relative scarcities and demands of alternative resources have directed the evolution of institutions in ways that have increased the productivity with which individuals use resources. They contend, furthermore, that there need not be a tradeoff between the increased total benefits resulting from institutional changes that permit more efficient use of resources and negative changes in the distribution of income among individuals in the economy. That is, institutional changes may provide the means to increase the total size of the pie and, as a result, everyone may receive a larger slice. In this framework, institutional innovation is seen as a process that can promote more rapid technical changes and reduce the negative economic impacts of high transactions costs, missing markets, and resulting externalities. In their discussion of technical change in agriculture, Hayami and Ruttan place great emphasis on the important role of nonmarket institutions — government agencies, national and international research centers, universities, and private firms — in the process of generating new agricultural technologies. Successful development of new technologies occurs when these institutions undertake activities to overcome the constraints of scarce factors on total production.

On the other hand, a different school of thought commonly referred to as the public-choice approach considers most nonmarket resource allocation institutions, including government activities, to be essentially wasteful (Batie 1984). Proponents recognize that there are certain areas of market failure that necessitate some sort of institutional response, such as government regulation. All institutions concerned with allocation of resources within a society are, however, assumed to have the implicit or explicit objective of increasing the welfare of particular interest groups rather than society as a whole. At the extreme, perfectly competitive markets consist of individual producers and consumers all pursuing their own self-interests. Neoclassical welfare theory describes conditions under which this type of institution maximizes total income of the economy. Coalitions, or organized interest groups, are also assumed to be competing with each other for the limited amount of economic profits (rents) available in the economy. Actions by these interest groups may distort the operation of competitive markets by imposing taxes or providing subsidies to certain activities to reduce aggregate income (Bhagwati 1982; Krueger 1974). Those groups able to marshal the greatest amount of political influence will be able to capture the largest share of the economic benefits.

Mancur Olson (1982) has proposed a dynamic extension of the public-choice view. He argues that broad-based coalitions of individuals in a society are needed in order to adopt institutional innovations that will provide benefits to a large share of the population. These large coalitions are harder to forge than more narrowly focused interest groups. Over time, interest groups proliferate and evolve institutions that stifle innovation and overall growth. Some sort of external shock, such as opening up of markets to free trade, is needed to undermine the control of narrow interest groups. Thus, the public-choice approach, in contrast to the induced innovation model of Hayami and Ruttan, considers institutional change to be motivated by attempts to increase the share of rents going to particular groups, and not necessarily driven to mitigate relative scarcities in the economy or to try to reduce transactions costs. Coalitions of individuals attempt to increase their share of the pie, but their actions tend to diminish the size of the pie, or in a dynamic world, diminish the rate at which the pie expands.

The policy implications of the public-choice paradigm are very different from those of the induced-innovation model. In the view of Hayami and Ruttan, many different types of institutions play important roles in the process of economic development. The goal of policies should be to ensure that these institutions have the flexibility to respond to economic signals of underlying scarcities in the economy. The public-choice model argues that institutions have an inherent tendency to distort resource allocations away from the efficient competitive market equilibrium, and, furthermore, tend to concentrate returns in the hands of the politically powerful. To the extent possible, the nonmarket institutions that govern resource allocations must be replaced with market-like institutions so that these unwanted tendencies will be minimized. The economic role of the government is to provide the necessary infrastructures that will permit the functioning of competitive markets.

The theoretical arguments just outlined are usually discussed in the context of national-level institutions and policies, although the logic can be just as easily applied to local-level institutions. One important difference in this context, however, is that national institutions can be taken as exogenous factors that have the potential to directly influence the behavior of local institutions. The external institutions may be either beneficial or detrimental to local populations. On the one hand, higher-level institutions may have the necessary authority and financial resources to successfully undertake public infrastructure investments—roads, educational facilities, large-scale irrigation networks, and so on—that are beyond the capacity of purely local groups to implement. On the other hand, outside institutions may restrict the flexibility of local institutions to address local problems. National government institutions throughout the world are notorious for being unresponsive to local concerns. Even worse, they are typically driven by political forces that are independent of, and possibly antagonistic toward, local community interests.

The induced-innovation and public-choice models each provide insights into possible strengths and weaknesses of local institutions with regard to economic efficiency of resource utilization and distribution of returns among users. Indigenous institutions are autonomous and therefore have great flexibility to take advantage of changing economic realities, as suggested by Hayami and Ruttan. Following the public-choice logic, however, we may expect that these changes will occur in such a manner as to ensure that those with the greatest political power receive a significant, if not exclusive, portion of the benefits to be gained from institutional innovations. Individuals possessing greater political power will be able to block any changes in which they are made worse off than under the status quo. On the other hand, they will support any changes that benefit themselves, even if these changes provide few gains to other members with less political influence. In these situations of unequal distributions of political power among individuals within local decision-making groups, there may be a need for a higher-level authority to intervene to ensure a more equitable distribution of benefits among individuals.

Many empirical studies of irrigation networks have addressed the relationship between management institutions and productivity of the networks. In particular, to what extent do the existing institutions permit or constrain the maximization of net economic benefits of the irrigation network as a whole, or of individual members within the network? If existing institutions do constrain productivity, can potential improvements be identified?[1] Much of this research indicates that existing institutional mechanisms do in fact constrain the available water resources from being used to their maximum economic potential by at least some individual users (see Maass and Anderson 1978; Skold, El Shinnawi, and Nasr 1984; Pasternak 1968; Yaron and Ratner 1990; Easter 1975; Reidinger 1974).[2] One common theme in these empirical studies of irrigation networks is the importance of distribution of political and economic power among network members (e.g., Hunt and Hunt 1976; Pasternak 1968). Wealth and incomes are usually very unequally distributed among members of irrigation networks. In many instances, larger landowners or farmers with land closer to the water source receive disproportionate amounts of water and bear less of the costs of adjusting to restrictions in overall water availability.

In the remainder of this chapter we examine the economic performance of small-scale irrigation networks in Cape Verde. These networks were installed and are nominally operated by a national agency, but in fact, within general operational regulations, local farmers undertake most of the day-to-day management of the system. Recent economic changes in Cape Verde have created strong pressures for change in irrigated agriculture. The degree and manner in which existing irrigation management institutions have permitted response to these economic pressures will be analyzed in relation to the theoretical descriptions of institutional change already outlined.

Irrigation Institutions in Cape Verde

Agricultural institutions in the Cape Verde Islands have undergone significant changes, particularly since the country's independence from Portugal in 1975. In addition, new technologies, rapid population growth, and changes in market demands for agricultural commodities have created incentives for institutional changes as described by Hayami and

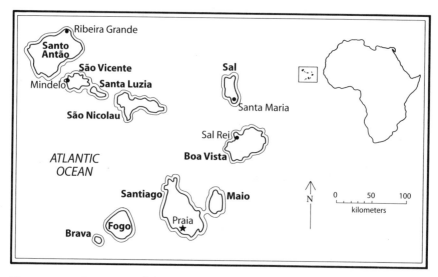

Figure 8.1 Location of the Cape Verde Islands.

Ruttan (1985). In this chapter we examine these economic pressures for change, and look at ways that local irrigation management institutions have adapted to these pressures.

Located 600 kilometers west of Senegal in the Atlantic Ocean, the Cape Verde archipelago (figure 8.1) falls within the Sahelian weather system. Average annual rainfall is less than 350 millimeters (Freeman et al. 1978). Rainfall patterns exhibit long-run cyclical patterns, with extended periods of drought. Uninhabited when discovered by Portuguese explorers in 1460, the islands were populated by Portuguese settlers who were granted land by the crown, and Africans who were brought to the islands in the growing slave trade, for which Cape Verde served as an entrepot. Most agricultural land in Cape Verde is rainfed. But rainfall is generally very scarce, and subject to significant year-to-year fluctuations; drought conditions lasting for many years are common in the historical record. Under these conditions, the productivity of rainfed agriculture is extremely low. The major rainfed crops are corn interplanted with beans; some restricted areas with more adequate access to moisture are planted in potatoes and peanuts.

The very low returns provided from rainfed agriculture place a high premium on irrigated land; however, the total amount of irrigated land, 2,990 hectares, represents just 7 percent of all agricultural land. In the country as a whole, only 23 percent of all farm households have access to any irrigated land. The average amount of irrigated land farmed by these households is less than one-tenth of a hectare (Ministry of Rural Development and Fishing 1990). The major irrigated crops are sugar cane, bananas, cassava, and sweet potatoes. Vegetables are grown in a few areas. Irrigated agriculture is practiced along *ribeiras,* steep valleys that run from the higher elevations of the islands to the coast. Traditional irrigation networks draw water from shallow hand-dug wells near the coast, or from natural springs found at the higher elevations of the ribeiras. More recently, tubewells have been drilled into deeper alluvial flows and fossil aquifers. Almost all irrigation installations in Cape Verde provide water to several farmers. The water distribution systems of these networks — usually plastic tubing or cement canals, although earthen trenches are still used in some areas — have very low flow rates. The traditional and more modern systems can only deliver water to one field at a time. Because only one network member is able to receive water at a particular time, farmers within a single network compete among themselves for their timing of access to water and to determine the total allocation over the entire growing season. Scheduling water deliveries to members is, therefore, a crucial management issue for the small-scale irrigation networks found in Cape Verde.

During colonial times, water sources were developed either by individuals or by the colonial agency charged with managing water resources on the islands. Individuals and groups of neighboring farmers built reservoirs near natural springs or dug open wells into the alluvium. Larger landowners, and in some cases the colonial agency, drilled deeper tubewells. The management of all these irrigation networks remained de facto in the hands of the users. In many instances, private wells were installed by large landowners who had access to capital necessary to finance such investments. These owners had first claim to the water, and the other users, often tenants of the well-owner, had access only to the surplus water that the owner did not use.

Cooperatively managed networks adopted a free-access rotation scheme, known as the *entornador-entornador* system. This system was also practiced under the control of the colonial agency. Each member could draw as much water as desired during his turn in the rotation. Under this allocation mechanism, the interval between irrigations for each network member depended on the amount of time all the other users in the sys-

Table 8.1 Percentage of irrigated area in Cape Verde, by crop, various estimates.

	Source of estimates		
Crop	Freeman et al. 1976	SCET/AGRI 1985	1988 Agricultural Census
Sugar cane	61	56	43
Tubers	13	14	27
Bananas	9	10	15
Vegetables	11	14	11

Sources: Freeman et al. 1976; SCET/AGRI 1985; Ministry of Rural Development and Fishing 1990

tem drew water. Each member, facing uncertainty about when he would next receive water, had strong incentives to put as much water as possible on his fields, hoping to retain moisture in the root zone as long as possible. In order to maximize the amount of water that could be stored in the soil, large rectangular basins were dug into the fields, which could be filled with water during irrigations. These basins provide a means for farmers to store water in their fields. To the extent that each member increased his access time to draw as much water as possible during his turn in the rotation, the amount of time between rotations was prolonged for all network members.

Agricultural practices have adapted to these constraints on access to water imposed by the limited physical capacity of the irrigation systems and by members' lack of control over their irrigation schedules. Traditionally, the predominant irrigated crops in Cape Verde have been sugar cane, cassava, and sweet potatoes, with localized production of bananas and vegetables in areas with more abundant water supplies.[3] Table 8.1 shows several estimates of the proportion of land planted with each of these crops. There are substantial differences among the individual estimates of cropping patterns, because they are based on different data sources, although all the estimates indicate the relative importance of sugar cane and tuber crops in overall production. The concentration in area planted to these crops, which are quite drought-resistant, reflects adaptation by farmers to the constraints they face with respect to access to irrigation water.

Sugar cane is used to make *grogue,* a type of rum widely consumed throughout Cape Verde. Although sugar cane requires much water over the year, it is able to withstand long periods between irrigations. In fact, growers often deliberately subject their mature crops to water stress in order to increase the sugar content in the stalks. In recent times, cane fields have often gone for periods of two months or more between irrigations (Finan and Belknap 1984; Langworthy et al. 1986). Sweet potatoes and cassava are usually intercropped. They are primarily grown for consumption by the household. These crops are also relatively resistant to long dry periods, being able to produce usable tubers even with irrigation intervals of up to one month. The yields of sweet potato, and especially cassava, increase significantly, however, if they receive water more frequently. There is a significant demand for tubers in urban areas and they command attractive prices at the farm gate, so they can be grown as cash crops if sufficient water is available.

The water requirements of bananas are more stringent than those of sugar cane or tubers. Farmers report that bananas should be irrigated at least every two weeks to provide the best yields, although some marketable production can be expected even with three-week intervals. During colonial times, bananas were grown to be shipped to Portugal. This was a profitable activity, and production was quite widespread up until the mid-1960s, the beginning of the most recent cycle of drought. Before that time, the area planted in bananas on both Santiago and Santo Antão was quite large. Santo Antão in particular, with abundant supplies of irrigation water, was quite specialized in banana production. The onset of the drought severely reduced water supplies on the island and led to large-scale substitution of land into more drought-tolerant crops, particularly sugar cane. Ribeiras that are now entirely planted in cane were major banana-producing areas thirty years ago. Santo Antão thereafter became famous as a major producer of grogue. On Santiago the shift out of bananas has been somewhat less pronounced, because the major banana-producing areas have been relatively less hurt by the drought. Production on Santiago is concentrated in a few specific areas where large farmers have their own deep wells that provide steady supplies of water throughout the year. These wells draw from fossil aquifers that have not been affected by the lack of rainfall in recent years.

Vegetables require water even more frequently than bananas. These crops can withstand intervals between irrigations of only one to two weeks before water stress seriously reduces production. As a result, only a very restricted number of farmers have adequate conditions to be able to grow vegetables. As will be discussed in more detail later, market demand for vegetables has been very limited until recent years.

Most farmers with irrigated land have concentrated their production in drought-resistant crops—sugar cane and tuber crops. These crops have provided significantly higher profits than rainfed activity, because of strong demand for grogue by the local population and high prices for banana exports.[4] Until relatively recently, there was not a significant market demand for vegetables. Under these conditions, the scheduling of irrigation water through free-access rotations was not a severe constraint on producers' behavior. The long intervals between irrigations that characterize this allocation mechanism did not conflict with farmers' preferred cropping patterns.

In the past twenty years, however, Cape Verdean society has undergone widespread changes that, among other things, have caused significant shifts in the demands for irrigated crops. In particular, per-capita incomes have increased quite rapidly over this period. Foreign aid of all forms has increased dramatically since independence, and much of this aid has been injected into the economy as increases in the disposable income of households through public works projects and increased employment, at relatively high salaries, in the government bureaucracy. Also resulting from the increased foreign aid activities, the international community in the urban centers has grown substantially. Emigration to Europe, particularly the Netherlands, France, and, more recently, Italy, has also increased significantly in the past two decades (Freeman et al. 1978). The most common recent pattern has been for individuals to emigrate, leaving their families in Cape Verde, and sending remittances that enter the local economy. There has also been a large movement of local population from rural to urban areas. All these factors have created a significant increase in the urban demand for foodstuffs. The increase in incomes has also led to a shift in demand toward vegetables. Production of vegetables, including Irish potatoes, has increased dramatically since the mid-1960s to meet this growing, largely urban, demand (see table 8.2).

The growing demand for vegetables and tubers from urban markets translated into high farm gate prices. As shown in table 8.3, in the mid-1980s the farm-level prices of vegetables were significantly higher than their costs of production, implying very high rates of profit for these crops. The data summarized in table 8.3 demonstrates that farmers with irri-

Table 8.2   Production of major irrigated crops in Cape Verde, 1971–1974 to 1981–1985.

Period	Crops				
	Sugar cane	Sweet potato	Cassava	Bananas	Vegetables
1970/74	8,824	1,434	1,285	5,392	373
1975/80	12,800	1,967	4,121	6,160	2,905
1981/85	9,840	2,960	1,890	4,000	3,788

Source: Ministry of Agricultural Development and Fishing, Agricultural Statistics, various years

gated land had very strong economic incentives to shift away from sugar cane and low-yield (infrequently irrigated) tuber crops into production of vegetables, high-yield tuber crops, and bananas.

At the same time these changes in demand for irrigated crops were occurring, management of irrigation networks also changed. Since independence from Portugal, the National Agency for Groundwater Resources (INRH; Instituto Naçional de Recursos Hídricos) has been responsible for the installation of new wells, and the management of all new and existing public irrigation networks. The INRH has replaced the free-access rotation schemes with a fixed-access rotation mechanism, in which members are allocated a fixed amount of time during which they may receive water during their turn in the rotation. When the INRH assumed control of irrigation facilities, an engineer would interview all the existing users to identify the amount of land used by each, and what kinds of crops they grow on their plots. Time allocations were then assigned to individual members based on the amount of land they operated and the types of crops they normally grew. These access times that are awarded to members will be referred to as *nominal allocations.* As will be described later, the actual or effective irrigation time, during which water actually reaches the field, is shorter than the nominal allocation. The pump operator — an employee of the INRH and usually also a member of the network — oversees the distribution of water to members. He ensures that members receive their time allotments in the rotation. He records the water deliveries to each member, and the INRH charges a nominal fee of one Cape Verdean *escudo* (71 CVE per dollar) per cubic meter based on these records.

The fixed-access rotation provides more control over the timing of irrigations for all the network members. Under this system, the interval

Table 8.3 Yields, prices, production costs, and net revenues for major irrigated crops in Cape Verde.

Crop	Irrigation Interval (days)	Yield (kg/ha)	Gross Price (CVE/kg)	Production Revenue ('000 CVE/ha)	Net Costs ('000 CVE/ha)	Revenue ('000 CVE/ha)
Sugar cane/ straw	30	1,500 300	200 80	324	144	180
Sweet potato/ cassava	30	3,000 5,000	35 40	305	49	256
Sweet potato/ cassava	15	9,000 15,000	35 40	915	64	851
Bananas	15	70,000	15	1,050	123	927
Potatoes	7	17,000	60	1,020	150	870
Tomatoes	7	34,000	45	1,530	186	1,300

Source: Authors' calculations

between irrigations for any member equals the amount of time required to make one cycle through the rotation. Assuming no down time between field applications, a fixed number of hours of operation per week, full use of allotted time by all members, and no stoppage due to breakdowns, each member would receive water on the same interval, described by the following equation:

$$I = A_j / (HR \cdot ND) \tag{1}$$

In this equation, I is the weekly interval between rotations, A_j is the hourly allocation of member j per rotation, HR is hours of pump operation per day, and ND is number of working days per week.

Under the fixed-access procedure, the INRH pump operator has nominal control over all the variables on the right-hand side of equation 1. In practice, however, the parameters HD and ND may change over time. There is substantial variation in the hours of operation per week because of equipment breakdowns, delays in moving the tubing, absence of the operator, and so on. These factors may cause I to vary over the course of the irrigation season, but the irrigation intervals of all network members would adjust identically, and all members should receive the same total number of irrigations over the season.

Detailed irrigation calendars obtained from farmers within an irrigation network in Ribeira Seca, a region of Santiago Island, reveal that, in fact, there is substantial variation in irrigation intervals among members within a single network. The time allocations for individual members established by INRH are closely correlated with total land area farmed, but the amount of water allocated per unit of land decreases as total land area increases. Thus, although larger landowners receive more total water allocation than their neighbors with less land, their water allocation per unit of land is less.

The expected time required to complete a rotation through the allocations of all members within this network was approximately three weeks under normal operating conditions — equivalent to twelve total irrigations over the 244-day period between January and the end of August. The actual records of irrigations during the 1989 irrigation season show substantial variation across the thirty-five farmers in the sample, ranging from two to thirty-four irrigations between January and August. In table 8.4, the surveyed farmers are divided into five groups according to the num-

Table 8.4 Number of farms, irrigated area, and water allocations in surveyed irrigation network, by number of irrigations per season.

Farm Group	Number of Irrigations	Average Number of Irrigations per Season	Average Irrigation Interval (weeks)	Number of Farms	Total Land Area (ha)	Average Farm Size (ha)	Total Water Deliveries (hrs/season)	Average Water per Land Unit (hrs/ha/season)
I	< 8	5	6	20 (57%)	1.99 (32%)	0.10	282 (24%)	17.8
II	8–11	9	4	6 (17%)	1.25 (20%)	0.21	184 (16%)	17.3
III	12–17	13	3	4 (11%)	1.08 (17%)	0.27	141 (12%)	21.4
IV	18–31	23	2	3 (9%)	0.84 (13%)	0.28	224 (19%)	27.8
V	> 31	33	1	2 (6%)	1.12 (18%)	0.56	329 (28%)	29.4
Total Sample		10	3	35 (100%)	6.28 (100%)	0.18	1,160 (100%)	19.7

ber of irrigations received during the irrigation season. At most, the first group irrigated seven times during the irrigation season, which corresponds to an average irrigation interval of nearly five weeks. The average interval between irrigations for the twenty farms in the first group was forty-nine days. For the remaining four groups, the number of irrigations corresponds to average intervals of approximately four weeks to one week, respectively.

This discrepancy between the expected irrigation interval, which would in principal obtain for all members, and the actual irrigation patterns of individuals, is due to two factors. First, some members do not actually use their allocations—either because they are not present at the scheduled time, they have no crops on their land, or they are in arrears on payments for previous water use and were therefore prohibited from receiving any more water. Other members may obtain these available hours, and thereby receive water more than once within a single rotation. Second, members may divide their time allotment into smaller segments and trade these with other members to obtain multiple access times within a rotation. These are bilateral trades among network members; the pump operator will comply with changes in the official rotation, but only on the agreement of both parties, and members must work out the trades themselves.

Information from the Ribeira Seca sample reveals that farmers with larger hourly water allocations per rotation generally received water more often over the course of the irrigation season. Equation 2 reports results of regression of number of irrigations against water allocations per rotation:

Number of irrigations = 6.32 + 0.07 · (Nominal water allocation per rotation)^2

(5.53)   (4.73)

$R^2 = 0.42$ $F = 22.40$

(Significant at the 0.001 level)   (2)

Numbers in parentheses above the coefficients are t-values.

Because total water allocation is based in part on farm size, large farms generally receive water more frequently than small farms. Table 8.4 reveals the relationship between farm size and access to water. The distribution of land and water is highly skewed among these subgroups. The two farmers who received thirty-two or more irrigations (group v) represent less than

6 percent of the sample, but they farm 18 percent of the total land and receive almost 30 percent of the water.

Several factors help to explain the positive correlation between members' hourly water allocations and the number of times they irrigate. First, individuals with larger allocations can be expected to be able to arrange more trades simply because they command a larger portion of the total supply of access time to be traded. As an extreme example, suppose a network has only three members, two with two-hour allocations and one with four hours, and that each member wishes to trade away half of his allocation so as to receive (at least) two irrigations during the rotation. The larger farmer controls half of the total time allotment for the rotation. If the exchanges were purely random, the large farmer would be twice as likely to be part of any exchange as either of the two smaller members. In fact, the most likely outcome would be for each of the two smaller farmers to trade one hour each with the larger member.[5] In this instance, the larger farmer gets three irrigations within the rotation because he has a larger share of the total nominal irrigation time for which other members are competing.

Second, because of transaction costs of trading time allocations with other members—search costs of identifying trading partners and risks inherent in agreements of asynchronous exchange—individuals will seek to minimize the number of transactions necessary to meet their desired reallocations. This fact also benefits members with larger nominal allocations, because they will have an advantage in finding trading partners. If a member wishes to trade two hours, he will prefer to trade with a single individual able to provide the full two hours, rather than make multiple trades to get the two hours from several other members. Thus, network members with larger hourly allocations have greater incentives to trade, and are likely to trade with other individuals with large allocations.

In addition, individuals with larger water allocations have relatively lower costs of splitting their allocations into several separate irrigations because of a fixed down time cost associated with each irrigation. Within the time allocated to a network member, the pump must be stopped, the tubing detached from the wellhead and that of the current recipient attached, fuel measured out and added to the tank, and after the pump is re-engaged, the tube must be filled before the water actually reaches the field. The down time cost is the difference between a member's nominal time allocation and the effective irrigation time during which his land actually

Table 8.5 Ratio of effective to nominal irrigation time, by number of irrigations per rotation (assuming 15 minutes of downtime per irrigation).

Water allocations	Number of irrigations per rotation		
	1	2	3
60-minute allocation per rotation			
Nominal allocation per irrigation (minutes)	60	30	20
Effective allocation per irrigation (minutes)	45	15	5
Effective/Nominal Ratio (%)	75	50	25
120-minute allocation per rotation			
Nominal allocation per irrigation (minutes)	120	60	40
Effective allocation per irrigation (minutes)	105	45	25
Effective/Nominal Ratio (%)	88	75	63
180-minute allocation per rotation			
Nominal allocation per irrigation (minutes)	180	90	60
Effective allocation per irrigation (minutes)	165	75	45
Effective/Nominal Ratio (%)	92	83	75

Note: Assuming 15 minutes of downtime per irrigation.

receives water. The length of down time varies according to the distance of the field from the wellhead, usually between ten and fifteen minutes. Down time is not proportional to the nominal allocation, however, so the larger the nominal allocation, the smaller is the down time cost per irrigation as a proportion of total effective irrigation time. The proportional cost of spreading a large time allocation into several different irrigations within a rotation is less than for a smaller allocation.

Table 8.5 shows the relationship between the size of the initial nominal allocation and the percentage decrease in effective irrigation time from splitting the nominal allocation into two or three separate irrigations. In this example, an initial allocation of one hour per allocation will provide an effective irrigation time of forty-five minutes, if water is applied once (assuming 15 minutes of down time per irrigation). If split into three separate irrigations, the effective time per irrigation is only five minutes, so the amount of time that water is actually going onto the field is only one-

Table 8.6 Area planted by crop type and farm category in surveyed irrigation network, by number of irrigations per season.

Farm category	Crop				Total land (ha)
	Sugar cane	Sweet potato/ cassava	Vegetables and bananas	Other	
I	0.52	0.87	0.60	0.00	1.99
II	0.32	0.40	0.47	0.06	1.25
III	0.50	0.08	0.23	0.28	1.09
Subtotal	1.34	1.35	1.30	0.34	4.33
% Total area	31	31	30	8	100
IV	0.05	0.12	0.57	0.09	0.83
V	0.12	0.26	0.63	0.11	1.12
Subtotal	0.17	0.38	1.20	0.20	1.95
% Total area	9	19	62	10	100
Total sample	1.51	1.73	2.50	0.54	6.28
% Total area	24	28	40	9	100

Source: Authors' Ribeira Seca irrigation network survey

quarter of the nominal allocation. With a larger allocation, the reduction in the effective irrigation time resulting from splitting the allocation into several irrigations is less. For example, splitting a nominal allocation of three hours into three different irrigations still permits 75 percent of the nominal allocation to be realized as effective watering time.

Variations in access to water have significant impacts on agricultural practices. Observed cropping patterns of the surveyed farmers in Ribeira Seca correlate closely with the number of irrigations they receive over the dry season. Table 8.6 shows the areas planted in drought-resistant (sugar cane), subsistence (sweet potatoes and cassava), and cash crops (bananas and vegetables) by categories of farms with different numbers of irrigations. There is a distinct difference in production patterns between those farms that receive water on an average interval of two weeks or less (Groups IV and V) and those with longer intervals (Groups I to III). Over 60 percent of the land of farms with intervals over two weeks is planted in the more drought-tolerant crops—sugarcane, sweet potato, and cassava. For the farms with intervals of two weeks and under, the share in these crops falls to 30 percent, and bananas and vegetables account for over 60 percent of cropped area. Thus, larger farmers received greater allocations of time in the rotation, and have been able to transact for more frequent access to water. This in turn has enabled them to grow more profitable crops, particularly vegetables, high-yield cassava, and bananas.

Policy Implications

The local institutional innovation of bilateral exchanges in water allocations has permitted some farmers in the Ribeira Seca irrigation network to grow crops with higher returns and more restrictive water demands than the traditional irrigated crops. Several factors have limited the benefits of this innovation to those members with large nominal water allocations (the larger farmers within the network). As the external agency charged with the construction, maintenance, and operation of the irrigation system, the INRH is in a position to alter its existing fixed-rotation scheme to provide the benefits of more frequent access to water for all members. The factors that limit the participation of the smaller members are their smaller allocations of time with which to trade, and the higher proportional down time costs they face from splitting up their allocations into multiple irrigations. An alternative scheduling regime could address these problems.

The simplest strategy would be to reduce all members' allocations by the same proportions, thereby reducing the expected interval between rotations. In fact, this strategy was already being discussed by the Ribeira Seca network at the time of our field research—specifically, reducing each member's allocation by one-third in order to cut the time between rotations from three weeks to two weeks. Although this strategy provides all members with water on shorter intervals, it does not permit flexibility for individual members to tailor their irrigation schedules to their own needs. Some farmers may not wish to grow the higher-valued crops, which also demand much higher levels of day-to-day management, and expose the producer to higher levels of production and price risks. On the other hand, those farmers wishing to grow vegetables will still need to get water more frequently, even with the shortened rotation interval. A scheduling scheme that can provide more flexibility to individuals is needed.

A more flexible strategy would be for each member to receive a total time allocation for the season (based on his share in the total land area of the network), and to devise a calendar for receiving the allocation over the course of the season. The pump operator would be responsible for devising a weekly irrigation schedule that would best meet the needs of all members. This scheme offers the advantage of permitting members to choose different irrigation schedules; those wishing to grow vegetables could receive water every few days, perhaps concentrated in two or three months, while growers of sugar cane could spread their irrigations over longer intervals. In addition, the operator has the flexibility to irrigate nearby fields sequentially, thereby minimizing down time costs. This mechanism would not provide flexibility for changes within the season.

A better solution would be for the operator to set up a weekly schedule. All members would indicate their desired irrigation plans for the coming week, and the operator would then design a schedule. If the demands of all members could not be met because of insufficient capacity, the schedules of all members could be adjusted proportionately, so all would bear the cost equally. In addition, the operator could design the schedule to cover adjacent fields sequentially, thereby minimizing the down time cost borne by the network as a whole. Under this scheme there would need to be some means of limiting individuals' weekly demands, otherwise everyone would have incentives to place larger requests so as to ensure receiving the desired amount, even with cutbacks. One way would be to place an upper weekly limit per farmer based on land area. This would produce an unequal distribution of water access among members, as in the present situation. Alternatively, a price could be charged for water that would discourage overuse. Then the issue of at what level the price should be set would have to be addressed.

Changes in Cape Verdean Irrigation Institutions

Over the past twenty years, increased demand for vegetables in Cape Verde has created growing economic incentives for institutional innovations that could provide water to farmers more frequently. At the same time, a combination of national and local changes have led to increased flexibility. At the national level, a state agency implemented a fixed-access rotation scheme that provides more control over the timing of water deliveries to individual farmers. Within this framework, farmers are able to undertake, and have undertaken, trades among themselves in order to readjust the timing of their water deliveries. A number of factors constrain smaller farmers from participating in such trades, and conversely, larger farmers are able to benefit disproportionately from these opportunities.

The observed changes in irrigation management institutions in Cape Verde suggest a process of change that has elements of both the induced-innovation model and the public-choice framework. On the one hand, there have been significant institutional changes within the irrigation networks in response to changing economic conditions. These changes have increased the ability of some producers to take advantage of new profitable opportunities. On the other hand, the beneficiaries of these institutional changes have been limited to larger farmers. Farmers with more land received a larger time allocation in the rotation devised by the INRH. This greater allocation in turn provided the larger farmers with more possibilities to make bilateral trades and thereby control more directly the timing of their access to water. With more frequent access to water, the larger farmers have been able to plant proportionately more of their land in higher-valued crops.

The example from Cape Verde of institutional changes in a small-scale irrigation network reveals the potential for locally directed institutional changes to favor those individuals already possessing greater economic and political power. Many empirical studies have found that larger and wealthier farmers have lower opportunity costs of acquiring many kinds of resources, in large part due to the nature of the institutions that govern the access to these resources (Berry and Cline 1979). This finding is supported in Cape Verde, where larger farms have greater access to irrigation water. In situations where unequal access to resources is tied to political power, institutional innovations to changes in economic conditions can be expected to channel benefits disproportionately to wealthier individuals. This may be appropriate from the point of view of technical (and even economic) efficiency. If, however, society is concerned with the equitable distribution of benefits, there is an explicit role for government to direct local institutions, namely to offset the skewed distribution of political power that can be found in local institutions. In the specific example of small-scale irrigation networks in Cape Verde described in this chapter, the government irrigation authority can implement alternative irrigation management schemes that provide smaller farmers with a more equitable access to water than the existing system.

Notes

1. See Bromley, Taylor, and Parker (1980) for a review of the literature.

2. Young, Daubert, and Morel-Seytoux (1986) provide an analysis of changes in water legislation that do improve the efficiency of water use relative to previous institutions.

3. Over the long history of the islands, production has shifted among a number of crops in response to long-term climatic shifts and changes in demand patterns from the outside world (Freeman et al. 1978).

4. Immediately after independence, Portugal reduced the level of purchases of bananas from Cape Verde; the level of shipments has increased significantly in recent years.

5. A bilateral trade between the two smaller members would leave the larger farmer without the possibility to split up his nominal allocation. Assuming positive benefits from increasing the number of irrigations, he would be willing to provide a more attractive trade (more than one-for-one exchange of irrigation time, or cash side-payment) in order to split his nominal allocation, so each of the smaller members could be enticed to trade with him rather than between themselves.

References Cited

Batie, S.
   1984   Alternative Views of Property Rights: Implications for Agricultural Use of Natural Resources. *American Journal of Agricultural Economics* xx: 814–18.

Berry, A. R., and W. R. Cline
   1979   *Agrarian Structure and Productivity in Developing Countries*. Baltimore: Johns Hopkins University Press.

Bhagwati, J. N.
   1982   Directly-Unproductive Profit-Seeking (DUP) Activities: A Welfare-Theoretic Synthesis and Generalization. *Journal of Political Economics* 90: 988–1002.

Bromley, D. W., C. Taylor, and D. E. Parker
   1980   Water Reform and Economic Development: Institutional Aspects of Water Management in the Developing Countries. *Economic Development and Cultural Change* 28:365–87.

Davis, L., and D. C. North
   1971   *Institutional Change and American Economic Growth*. Cambridge: Cambridge University Press.

Easter, K. W.
   1975   Field Channels: A Key to Better Indian Irrigation. *Water Resources Research* 11: 389–92.

Finan, Timothy J., and John Belknap

1984   *Characteristics of Santiago Agriculture: 1984 Report on Survey of Santiago Agri-culture.* Praia: USAID/Food Crops Research Project.

Freeman, P. H. et al.

1978   *Cape Verde: Assessment of the Agricultural Sector.* McLean, Va.: General Research Corporation.

Hayami, Y., and V. R. Ruttan

1985   *Agricultural Development: An International Perspective.* Baltimore: Johns Hopkins University Press.

Hunt, R. C., and E. Hunt

1976   Canal Irrigation and Local Social Organization. *Current Anthropology* 17: 389–411.

Krueger, A. O.

1974   The Political Economy of the Rent-Seeking Society. *American Economic Review* 64:291–303.

Langworthy, Mark W., Timothy J. Finan, Raul Varela, Elísio Rodrigues

1986   *Characteristics of Santo Antão Agriculture.* Praia: USAID/Food Crops Research Project.

Langworthy, M. W., and G. L. Thompson

1991   *Scheduling Irrigation Water: Equity and Production Impacts of Alternative Irrigation Regimes in Cape Verde.* Department of Agricultural Economics Working Paper No. xxx. Tucson: University of Arizona.

Maass, Arthur, and Raymond L. Anderson

1978   *. . . and the Desert Shall Rejoice: Conflict, Growth and Justice in Arid Environments.* Cambridge, Mass.: MIT Press.

Ministry of Agricultural Development and Fishing

1990   *Agricultural Census, 1988.* Cape Verde: Praia.

Olson, M.

1982   *The Rise and Decline of Nations.* New Haven: Yale University Press.

Pasternak, B.

1968   Social Consequences of Equalizing Irrigation Access. *Human Organization* 27: 332–43.

Reidinger, R. B.

1974   Institutional Rationing of Canal Water in Northern India: Conflict between Traditional Patterns and Modern Needs. *Economic Development and Cultural Change* 23: 79–104.

SCET/AGRI

1985   *Ilha de Santiago: Monograph.* Paris: Ministère Francais des Relation Exterieurs Cooperation et Developpement.

Skold, M.D., A. A. El Shinnawi, and M. L. Nasr

1984   Irrigation Water Distribution Along Branch Canals in Egypt: Economic Effects. *Economic Development and Cultural Change* 32: 547–67.

United States Agency for International Development (USAID)

1987   Crop Budgets for Selected Irrigated Crops on Santiago and Santo Antão Islands. Mimeograph, Food Crops Research Project, São Jorge, Cape Verde.

Yaron, D., and A. Ratner

1990   Regional Cooperation in the Use of Irrigation Water: Efficiency and Income Distribution. *Agricultural Economics* 4:45–58.

Young, R. A., J. T. Daubert, and H. J. Morel-Seytoux

1986   Evaluating Institutional Alternatives for Managing an Interrelated Stream-Aquifer System. *American Journal of Agricultural Economics* 68:787–97.

Part 3

Development Lessons from Local Irrigation

9

Qanats and Rural Societies
Sustainable Agriculture and Irrigation Cultures in Contemporary Iran

Michael E. Bonine

Q*anats,* subterranean water conduits, were a fundamental part of rural society in many parts of the arid Middle East and Central Asia, but perhaps in no other region did they become so predominant as in the central plateau of Iran. Tens of thousands of qanats have enabled productive oases to dot the otherwise barren, desert landscape. Even in areas in which the water table lay deep underground, this traditional technology provided the means to irrigate and cultivate extensive fields as well as supplying ample amounts of water for even cities to flourish. This paper examines briefly the characteristics of the qanat as well as its role in Iranian rural society. It focuses on how water is managed and controlled locally and the implications of this organization. It indicates how these irrigation systems have influenced the morphology of settlements in Iran. The decline of some qanat systems is examined, particularly the problems related to the competition with pump wells, which usually lower the water table rather drastically. Finally, there is an assessment of the value of qanats as a locally managed and sustainable irrigation system.

Qanats in Iran: The Physical Basis

Within Iran dryland farming from rainfall is located mostly in the north and west, while irrigated farming—and qanats—occur principally in the central and eastern parts of the country (figure 9.1). The Elborz Mountains in the north and Zagros Mountains in the west provide rather effective barriers for rainfall to occur on the leeward sides, and so the annual

This chapter is an updated and extensively revised version of a paper, "Qanats and Irrigation Cultures in Iran," that was presented at The International Conference on Karez Irrigation, in Urumqi, China, in August 1990, and published in *The Proceedings of [the] International Conference on Karez Irrigation,* ed. International Conference on Karez Irrigation, 117–32 [English Section]. Urumqi, China: Xinjiang People's Publishing House, Educational and Cultural Press Ltd., 1993. Information on qanat irrigation practices is especially from Bonine 1982.

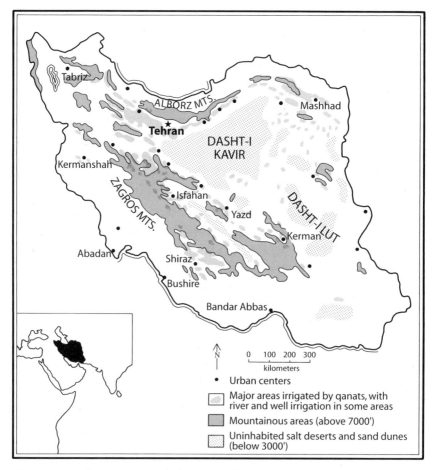

Figure 9.1 Qanat zones and irrigated areas in Iran. Many small, local qanat irrigated areas not shown, and wells have replaced qanats in some areas. Rainfed (dryland) farming occurs mainly in western and northern Iran, where considerable land is also irrigated by streams and rivers. Some rainfed farming occurs with qanat irrigation in the higher elevations, especially in wetter years (compiled from Ehlers 1980, 1984; Beaumont 1971; Bonine 1982, 1989; English 1966).

rainfall in the interior is less than 300 mm, generally decreasing to the east and south. Average annual rainfall totals in the interior of Iran are low; for instance, Isfahan receives 126 millimeters a year, Kerman 203 millimeters, and Yazd 67 millimeters. Yet, nearby local mountain systems and highlands receive considerably more rainfall (and even snow), which provides the majority of the water for the qanat systems. In the extreme southeast, as in the Dasht-i Lut, the rainfall is less than 15 millimeters a year, and with fewer highland areas there is less groundwater for qanats (Ehlers 1980).

The nature of qanats in Iran has been described in a number of publications (e.g., English 1968; Wulff 1968; Beaumont 1971, 1973, 1989; Goblot 1979; Behnia 1980s), so only a brief summary is presented here. A qanat is an almost horizontal, underground conduit that originates within an aquifer at a higher elevation, and the water flows in the qanat by gravity (figure 9.2). The slope must be very slight—not too steep or erosion will occur, or if there is not sufficient gradient, too much seepage occurs. In Iran qanats usually tap water bearing zones in alluvial fans at the base of mountains, which receive more moisture than the lower elevations.

The source of the qanat, the mother well, is sunk into the groundwater recharge zone, and this vertical shaft can be over 100 meters in depth— although it usually is less than 50 meters (Beaumont 1989:19). The length of the horizontal tunnel may be several hundred meters or it may be 10 to 20 kilometers or more. "The length of a qanat is controlled by the depth of the mother well and the slope of the ground surface. The qanat tunnel has to slope gently downhill for water to flow along it under the influence of gravity. The critical factor, therefore, becomes the slope of the ground surface that will eventually intersect the tunnel. This means that the shortest qanats occur where steep ground slopes are found and the longest where gentle surface slopes predominate" (Beaumont 1989:23). The point at which the qanat exits onto the surface usually is where cultivated fields and settlements are found.

In Iran, most qanats flow in direct response to precipitation, snow melt, and floods, reflecting the high permeability of the alluvial sediments in which groundwater recharge occurs. Hence, the flow becomes variable, changing with the rainfall patterns and the seasons. The qanats usually tap alluvial fans where material is coarse and quite permeable. Water originating in the alluvial fans is quite pure and salt-free (called "sweet water," or *qandab*). Qanats also flow at the maximum in late winter and early spring, the optimal period for watering winter wheat and barley as well as primary fruit crops such as apples and almonds. There is much less qanat water in the summer, during the growing season for melons and vegetables (figure 9.3). A few qanats tap very deep and less changeable water tables, and hence have constant flows throughout the year.

Beaumont (1971) has noted that measurements of 2,000 qanats in Iran

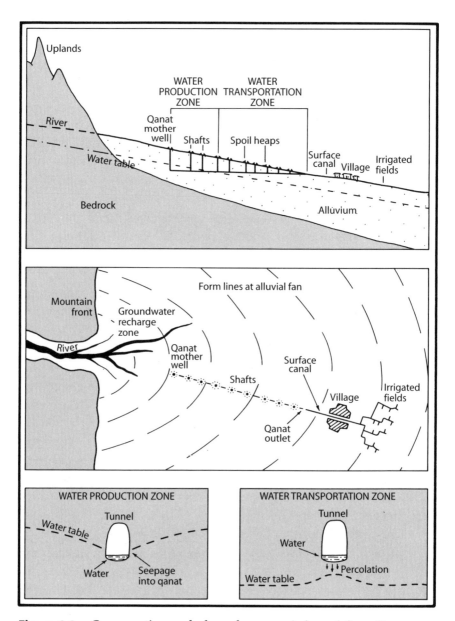

Figure 9.2 Cross-section and plan of a qanat (adapted from Beaumont 1973).

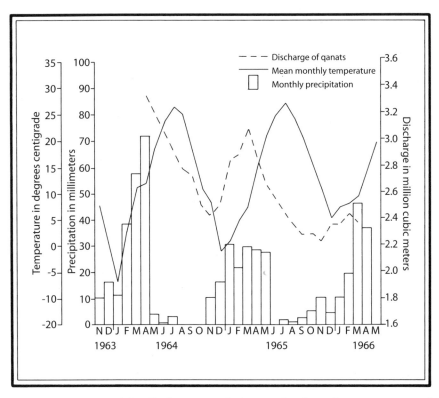

Figure 9.3   Monthly discharge variations of selected qanats around Mashhad, Iran (data from Beaumont 1971).

(in the 1960s) revealed variations in discharge from 0 to 300 cubic meters per hour, with two-thirds less than 60 cubic meters per hour and almost half (45 percent) with less than 30 cubic meters per hour. A typical qanat may irrigate only about 10 to 20 hectares, and in many cases much less — as few as 1 to 2 hectares. In situations where the flow of the qanat is very small, a diversion dam or pond is used to collect the water, and then the water is discharged at specific intervals.

The number of currently functioning qanats in Iran is difficult to determine, for within the last several decades of the Pahlavi regime (to 1978–1979) thousands (if not several tens of thousands) of qanats dried up due to drought, neglect, or falling water tables from competition from wells. Yet, under the Islamic Republic, with a new emphasis on local agriculture, qanat systems may have begun to increase in number. In any case, the estimates of the number (and flow) of qanats in use at any specific

Table 9.1 Qanats in Iran, 1954–1977.

Year	Annual Discharge (billion cubic meters)	Number of active qanats	Average Discharge of each qanat (liters/sec)
1954	18.10	21,060	27.25
1961	17.66	22,000	25.45
1963	16.00	20,000	25.35
1972	5.57	14,986	11.75
1973	6.23	15,500	12.75
1974	6.17	14,778	13.25
1975	6.95	15,770	13.97
1977	7.50	18,400	12.93

Source: Safinezhad 1356/1977–1978; Sadr 1357/1978–1979, quoted in Ghayoor 1990

time are subject to considerable error. In the early 1960s there were per-haps 20,000 to 22,000 qanats in Iran, declining to 15,000 to 18,000 by the late 1970s (see table 9.1). McLachlan (1988: 89) quotes official censuses of agricultural statistics for 1352 (1973–1974) that say there were 46,303 qanats still in use out of a total of 62,000. Official Islamic Republic statistics also state that there were 18,388 qanats in 1976–1977, and that the number had increased to 28,038 by 1985–1986 (statistics quoted in Schirazi 1993:283). More significant is the fact that the flow of many qanats has decreased considerably, from an average of about 27 liters per second per qanat in 1954 (or 25 liters per second in 1961 and 1963) to 12 to 13 liters per second in the 1970s (table 9.1).

The total annual discharge of qanats has fallen from around 16 to 18 billion cubic meters in the 1950s and early 1960s, to about 6 to 9 billion cubic meters in the 1970s (Sadr 1357/1978–1979, cited in Ghayoor 1990; McLachlan 1988:22). In 1977 the underground water resources in Iran pro-vided annually 18.5 billion cubic meters, of which the 18,400 active qanats provided 7.5 billion cubic meters; 16,600 deep wells furnished 7.5 billion cubic meters and 42,000 shallow wells 3.5 billion cubic meters. As men-tioned, the number of qanats had increased to 28,038 by 1985–1986, yet, the total quantity of water from qanats had increased to only 9.1 billion cubic meters (table 9.2) (Schirazi 1993:283). As a further comparison, Safi-nezhad (1977:24) notes that in 1973–1974 for all irrigated agriculture in

Table 9.2 Use of groundwater in Iran, 1976–1977 to 1985–1986.

Year[a]	Deep wells	Shallow wells	Qanats	Springs	Total
1976/77					
Number	16,626	42,546	18,388	8,193	85,753
Quantity[b]	7.47	3.88	7.54	5.45	24.34
1985/86					
Number	45,355	119,068	28,038	20,244	212,705
Quantity[b]	19.12	10.49	9.08	8.13	46.81

a. The Iranian year begins on March 21 of the western calendar year.
b. billion cubic meters.
Source: Adapted from Schirazi 1993:283, Table 12.5; Statistics from Markaz-i Amar (Center for Statistics), Islamic Republic of Iran 1982 and 1988

Iran, 17.7 billion cubic meters were provided annually by springs, qanats, and wells and 16.3 billion cubic meters by rivers (mostly in the north and west) — for a total of 34 billion cubic meters being utilized for irrigated agriculture. To put this in perspective, the 34 billion cubic meters is about 40 percent of the Nile's average annual flow of 84 billion cubic meters [measured at Aswan]; and so qanats in Iran in the 1950s supplied about 20 percent of such an amount, while today the water from qanats in Iran is about 10 percent of the Nile's average annual flow.

Only 9 to 10 percent of Iran's total area of 164.8 million hectares is arable, including fallow land. In the early 1970s, about 15 million hectares of land in Iran was under cultivation, including 2.5 million hectares of irrigated crops, 5.7 million hectares of rain fed crops, and 6.9 million hectares lying fallow. Other statistics (for 1974–1975, for instance), indicate that 16.4 million hectares were arable, of which 4.1 million hectares were irrigated, 6.4 million hectares were nonirrigated (rain fed), and 6 million hectares were fallow (table 9.3). Islamic Republic statistics indicate that the arable land had increased only slightly to 16.9 million hectares in 1988–1989 — in which there were 5.6 million hectares of irrigated lands, 5.9 million hectares of nonirrigated lands, and 5.4 million hectares of fallow lands (table 9.3). McLachlan (1988:100) notes that during the Pahlavi period no more than 225,000 hectares had come under new irrigation projects under the five development plans that ended in 1978. With an emphasis on small

Table 9.3 Arable land in Iran.

Year[a]	Arable	Tilled (% of arable)		Fallow (% of arable)		Irrigated (% of tilled)		Non-irrigated (% of tilled)	
1974–1975[a]	16.4	10.5	(64%)	6.0	(36%)	4.1	(39%)	6.4	(61%)
1978/1989	14.9	9.2	(62%)	5.7	(36%)	3.8	(42%)	5.4	(58%)
1982/1983	14.8	9.8	(67%)	4.9	(33%)	4.0	(40%)	5.9	(60%)
1985/1986[b]	18.5	12.2	(66%)	6.3	(34%)	4.3	(35%)	7.9	(65%)
1988/1989	16.9	11.5	(68%)	5.4	(32%)	5.6	(49%)	5.9	(51%)

a. The Iranian year begins on March 21 of the western calendar year.
b. Data from self-sufficiency plan.
Source: Adapted from Schirazi 1993: 281, Table 12.4; Statistics from Markaz-i Amar (Center for Statistics), Islamic Republic of Iran 1977 and 1990.

local systems during the Islamic Republic, of the 5.6 million hectares of irrigated lands in 1988–1989, it is probably true that at least 5 million hectares are being managed as local and small systems. It is also significant to note that groundwater use in Iran has almost doubled from 1976–1977 to 1985–1986, the major increase coming from shallow wells, deep wells, and even springs (table 9.2).

Qanat Construction in Iran

Qanats in Iran are dug by *muqanni,* professional diggers who have mastered this difficult, dangerous skill (Lambton 1989; Bonine 1982:145–48). Construction begins only after a master muqanni estimates where the qanat should originate and the exact route and grade of the conduit in order to exit at an existing or proposed village. One or more test wells are dug at the estimated water source to determine the depth and extent of the aquifer, and hence the depth of the conduit to be constructed. Digging of the conduit, however, begins where the qanat will exit (the *mazhar-i qanat).* As the qanat channel is being tunneled upslope toward the test well, vertical shafts must be dug for ventilation and to lift out the spoil. The distance between vertical shafts depends upon the depth of the shafts and the type of material being dug. Deeper shafts and less porous material require that they be closer together. The conduit is tunneled to the water table and then beyond to provide a sufficient water bearing zone for the qanat. This zone can be several kilometers in larger flowing qanats. The conduit may be dug only to the test well or it may end before or beyond that point. The deepest or head well is called the *pish-i kar* (before the channel) or the *madar-i chah* (mother well). Qanats, especially larger ones, may have a number of branches that tap the water table. These can be many kilometers in length, each having their own vertical shafts and mother well.

Qanats must be periodically cleaned and repaired; this is one of the main jobs of the muqanni. Mineral salt deposits accumulate on the channel bed of some qanats and the flow lessens if the deposits are not dug out every ten to fifteen years. More common is the need to repair or clean a qanat after a storm has poured rock and mud through the vertical shafts. Spoil heaps are left around the shafts in order to prevent this occurrence. If the water table drops, a common situation with the advent of modern deep water wells and pumps, the qanat channel must be extended and a new mother well dug. The mother wells of some qanats are extended every year. Many of the villages in Buluk-i Rustaq, an area northwest of Yazd, have their mother wells in the vicinity of Yazd where the water table has been dropping at least 1 meter per year for the past several decades. Many of these qanats have gone completely dry.

Qanat construction has always been very expensive, and only individuals with considerable capital could afford to finance such an enterprise (English 1966:139–49; McLachlan 1988:279fn 51). Hence, only large landlords and merchants, governors, and other wealthy or influential persons could afford to invest in a qanat. The great costs in constructing qanats was conducive to landlord control and promoted urban dominance by absentee landlords (Bobek 1974, 1979; Ehlers 1978; Bonine 1980b). Qanats also took many years to construct, and as Beaumont (1989:27) has noted, "there is a long period between the initial investment being made and any returns being obtained in the form of water production."

The cost of a qanat can be better understood with an example (Ehlers 1989:105). In 1979 in Assadabad (in western Iran in the Zagros region), a small qanat of 2 kilometers with a depth of the mother well of 16 meters and a total of fifty-seven vertical shafts cost on the average as follows:

1. shaft construction (200 tomans per m) – 456 m × 200 tomans = 91,200 tomans;
2. construction of the water channel (250 tomans per m) – 2,000 m × 250 tomans = 500,000 tomans; and
3. cementation of the qanat exit, the pond, etc. – approximately 20,000 tomans.

With an exchange rate of 7.6 tomans = $1.00, the total of 611,200 tomans is equivalent to $80,000. This is a very small and shallow qanat. Deep and long qanats would require much more money. Even in the 1950s and 1960s hundreds of thousands of dollars were required for a long and deep qanat, as in the Kerman region (English 1966:139–40). With inflation, if dug the traditional way, such qanats today would cost in the millions of dollars.

Qanat Irrigation:
The Division and Distribution of Water

Qanat water is used in a cycle that consists of a specific number of days (and nights) (Bonine 1982:148–52). In central Iran, for instance, individual village cycles range from six days to as long as twenty-two days. A farmer who owns or rents a share of qanat water is entitled to that share of water the entire year, not just for one specific cycle. Qanats with the same cycle can differ considerably in volume, and so the length of the cycle is not based upon the water flow. Soil conditions or the types of crops do not seem to be the reason either, and the length of cycles are often similar in specific regions, indicating perhaps a cultural-historical development (Bonine 1982).

Division of a water cycle into time shares is the basis for water distribution. A water share entitles the recipient to all the qanat water during that one time period. These units are not shares in terms of a specific volume of water, and so, if the volume of the qanat varies, so does the amount of water received during a specific time share. In the Yazd region of the central plateau the time shares range from a twenty-four-hour (*shabaneh ruz* = "night and day") or twelve-hour (*taq*) period for very small flowing qanats to units that are seven and a half minutes to twelve minutes each (often called *jurreh* or *sabu*) for the larger qanats. For such latter units, for every shabaneh ruz (twenty-four hours) there will be as many 128, 130, 140, or 144 jurreh. A village with 140 jurreh per twenty-four hours and a ten-day cycle would therefore have a total of 1,400 water shares to be used by the farmers for the entire cycle. (See Bonine 1982 for further explanations, examples, and variations.)

The smaller water shares traditionally were timed by using a water clock, a copper bowl or *tashteh* that has a small hole in the bottom. Placed on the surface of a larger water vessel, when the bowl sinks it is equal to the time of one water share. If a farmer owns or rents five shares of water, during his turn for water the bowl (operated by the *mirab*, the person who times the shares) will sink five times. Although watches and clocks have replaced most *tashteh,* numerous village qanats, especially more remote ones, were still using the water clock in central Iran in the early 1970s (Bonine 1982).

The oldest type of traditional irrigation in Iran (and the Middle East) is flood irrigation, also called border or strip irrigation. Whether from a qanat, river, or wells, water is conveyed in an open ditch system of channels (*jubs*). In central Iran the qanat water flows into a series of rectangular strips (*korts*) where low and parallel border ridges control the water, and crops are irrigated by inundation. The korts are constructed with their long axis in the direction of greatest slope and water is diverted into

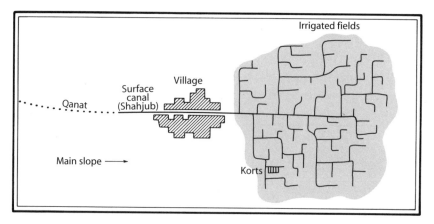

Figure 9.4 Rectangular channels and korts of an Iranian village (adapted from Bonine 1982).

the upper end of each strip, advancing down the slope in a thin sheet. The size of the kort depends upon the slope, flow of water, and the soil texture, and so sizes can vary considerably from village to village. For the irrigated lands of a village, considerable labor (and skill) is needed to design the system so that proper slope conditions will enable water to flow from the channels to the strips and even from kort to kort.

A grid pattern of channels on each side of the main channel (*shahjub*) distribute the water to the rectangular system of korts (figure 9.4). Hence, the border strips do not follow the contour, but the entire system is oriented the same way—the direction of greatest slope. Water is diverted from one channel to another or into fields merely by blocking the appropriate channels with mud dams, although wooden checkgates are used in a few large villages where water flow in the shahjub is quite swift. The predominately rectangular irrigation and field system not only facilitates the irrigation itself, but it also helps in measuring divisions for inheritance and selling. Because the qanat water is used in time shares, calculations on how long to water each specific plot are crucial. The more regular (rectangular) and similar in size, the easier it is to calculate the length of time needed for watering individual fields—which can be done quite easily by the local users. Because in most cases a qanat's flow is seasonal and varies due to the amount of rainfall, the number of fields that can be irrigated depends not only on the total number of water shares but also on the volume of the qanat's flow.

A core area of fields is always watered sometime during a complete cycle, and so it is more intensively used. Double-cropped fields and orchards, which may also contain crops, constitute this zone. These fields are usually the most rectangular and uniform. Beyond these parcels, usually downslope, are other plots that are used in a system of single cropping and fallow, their use depending upon the water supply. Such fields also can be watered seasonally by a river system, which may either be an entirely separate network of channels or feed into the qanat irrigation channel network. The fields watered only by a seasonal river tend to be more irregular because precision of timing is not as crucial. Such water is more variable in volume and comes suddenly, lasting only a short period. The more water, the more fields to be irrigated.

The overall morphology of the field system and the settlement are also influenced by the fact that an initial qanat irrigation system must be planned and constructed essentially as a single unit. When a qanat is built the flow is estimated, and then, depending on the flow and the designated cycle, the amount of land to be irrigated must be determined. Then the network of channels must be dug to reach the various parcels — which also must be prepared for the correct size and slope. Certainly, the construction of a qanat irrigation system was not only costly, but also most skillful and labor intensive.

Qanats and Irrigation Cultures in Iran

Qanats and qanat irrigation have had a profound influence on the life of the Iranian villager, for the cycle of water use, seasonal ebb and flow of the qanat, and rainy and drought periods all influence the agricultural and rural activities and "impose a rhythm on life" on the villagers (McLachlan 1988:93). In fact, Iranian qanat irrigation has produced what can be called "irrigation cultures" (McLachlan 1988), for we cannot understand the traditional village life in the central plateau of Iran without understanding the role of the qanat. As McLachlan (1988:90–91) notes:

> Reliance on the *qanat* by small and large landowners gave a powerful incentive for joint action in ensuring water supply, managing the irrigation cycle and preventing waste. It might be argued that for as long as the *qanat* remained the sole source of water it created and kept in being a strong element of social cohesion, even though this was ex-

pressed in many villages in oppressive relations between the more and less powerful participants in the irrigation system that it spawned.

Water from qanats is often called *ab-i deh* (water of the village), a term not used for water from wells or rivers. Villages are so connected with the water cycle of their qanats that when qanats have declined drastically in flow—or totally dried up—village social and economic relationships are changed, often rather drastically. The local villagers are directly involved in the distribution or overseeing of the qanat water (Lambton 1953; Bonine 1982:155–57). Although small qanats (and hence small villages) often do not have specific persons designated to supervise or oversee the division of water, the larger systems do require specific individuals to ensure their proper operation. Often one or two persons—the mirabs—are in charge of overseeing the timing of the shares, either by the water clock (bowl) or now, more commonly, by a watch. The mirab keeps a record of the use of everyone's water shares, which takes into account the trading, buying, and selling of water shares, and determines that each individual gets the correct number of shares (Bonine 1982). Other officials include persons to keep channels cleaned; individuals assigned to open and close certain channels, checkgates, or the pond that has collected water over a period of time (for smaller qanats); and persons to protect the cultivated fields and orchards from animals and thieves. Payment to these officials may be in kind or in a certain number of water shares per cycle.

Local agriculture in Iran traditionally (before land reform) was done by work teams of sharecropping peasants, called *buneh*s (Hooglund 1981, 1982). In fact, central and eastern Iran, the area where qanats occur, has been identified as the principal zone for these bunehs (Safinezhad 1977). The work teams often included the ownership of oxen (*gav*) for plough-ing, and so the teams were also called *gavband*. A village might have a dozen or so bunehs that were assigned specific lands to irrigate from the qanat. In some instances land was drawn by lottery and hence changed each year; in other villages, peasants who had the right to cultivate (pos-sessing the *nasaq*) kept the same lands. A buneh usually consisted of four to seven men, but it could be as few as three and as many as fourteen in larger villages. Each buneh was under the supervision of one peasant (the *sarbuneh*), who was appointed by the landlord. The teams worked together ploughing, sowing, irrigating, weeding, spreading manure, and harvesting.

Iranian land reform in the 1960s did away with sharecropping and many other inequities. As peasants who possessed the nasaq acquired land (and water), the buneh structure disappeared. For instance, in the small village of Talebabad (studied by Safinezhad) all sixty peasants were organized in fifteen four-man bunehs in 1965–1966 that were gone by 1975–1976; and thirty-nine of the sixty peasants worked alone (Hooglund 1981: 203). "The decline in importance of the *buneh*s was a direct, albeit unintended consequence of land reform" (Hooglund 1982:101). Individual peasant farmers, now interested in maximizing their own profits, have attempted to manage the irrigation and cultivation themselves without any major cooperation. Rural cooperatives have expanded their membership during the Islamic Republic, but their role for irrigation is not clear and often not effective, for "different factions within the government have opposing views on whether the cooperatives should function through local initiatives or be tightly controlled by the state" (Ajami 1993:257).

The key point is that qanat water in Iran has been (and is) locally managed, which also is verified by a number of village and ethnographic studies (e.g., Alberts 1963; Spooner 1974; Kielstra 1975, 1989; Hartl 1979, 1989). Especially before the 1960s land reform, the majority of villages and qanat water were owned by "landlords" (*arbab*) and not by the peasants (and buneh teams) who were working the land and using the water (see Lambton 1953). The arbab often were absentee landlords, living in the provincial capital or even Tehran. In other cases the landlords lived in the villages, although there were usually several important families who owned most of the water and rented it out to individual farmers (or bunehs).

In any case, even for the absentee-owned villages (before land reform), the irrigation system was managed entirely locally. The mirab and other officials were local villagers, and decisions for the distribution of the system were being made at the village level (although certainly affected by the landlords). Repairs and cleaning of the qanat (every several years) were handled by assessments on each share holder, often made annually and saved until needed. At least in the Yazd area, these assessments were taken out of the payment for each water share—and hence the landlords were theoretically paying for the upkeep of the qanats (which they owned).

Although there were certainly economic (and social) inequities in the traditional landlord-peasant relationships, the qanat irrigation systems in Iran were managed quite efficiently. This is emphasized by the fact that many qanats deteriorated during the 1960s when there was great uncertainty about the status of future ownership of land and water, so nobody wanted to invest in the upkeep of the qanats. Also, in many cases when the peasants who had been renting qanat water received the water (and land), they did not continue to invest in the qanats. The traditional system of qanat upkeep was no longer in place, and there was no effective new system. The farmers had no compulsion to invest in the long-term future of the qanat and its water. In some instances, newly established cooperatives have made funds available for the repair and upkeep of the village qanat, but this appears to be the exception rather than the rule. In fact, such cooperatives are more likely to promote and lend money for the drilling of wells, which may only lead to the further decline or even demise of qanats.

Qanat water is also an integral part of the socioreligious system in Iranian villages. Water from the channel is used to fill the tanks and fountains used for ablutions in the village mosques, and many shares of water are often *vaqf* or religiously endowed property. Hence, the rent for some shares of water is used to support a mosque or religious event, such as the Muharram ceremonies practiced by these Shi'i Muslims. Sometimes a major portion or even all of a water cycle is vaqf. The endowed water shares may be for institutions outside of the village as well, particularly for specific mosques or shrines of the provincial capital near the village and where the absentee landlords may live. Therefore, vaqf water in villages around Yazd, Kerman, or Kashan will sometimes support institutions in those towns. Needless to say, when such qanats decline or even dry up, there is much less support for specific local religious structures or for charitable purposes, impacting not only individuals but also the socioreligious institutions in this Islamic society.

Qanats and Iranian Settlement Form and Social Structure

Qanat irrigation and agricultural practices have also played an overriding role in the layout and morphology of traditional Iranian settlements (Bonine 1979, 1989). Most settlements in the central plateau exhibit a rather distinctive orthogonal or rectangular network of streets and houses. This grid pattern has resulted from the rectangular irrigation and field system. As villages (and even cities) expanded, they filled in adjacent rectangular fields and orchards. The orientation of the irrigation network (and fields) was determined by the need to arrange the rectangular fields and orchards to the slope of the land. Major streets and many blind alleys already existed within the field patterns before houses spread into these areas. Even the sizes and shapes of new houses are governed by the pre-existing system of fields and passageways (Bonine 1979).

The specific village site is determined by the qanat and the irrigation system. Because water is available only beginning at certain depths in alluvial fans, the exit of a slightly sloping underground channel can occur only at certain places. Once a village is established any subsequent new qanat must be constructed even more carefully and precisely, so that it exits at the appropriate part of the settlement and fields.

Clean drinking water must be available for inhabitants, so the exit of the qanat must be a proper distance upslope from the village and fields, although larger settlements that have many qanats do not have to use all of them for drinking water, so some systems can exit at fields even below the village. In some cases, as in the linear alluvial fan villages of the Kerman Basin, the main water channel actually passes through household compounds (English 1966). When qanats flow beneath a settlement (which can include qanats destined for other villages many kilometers downslope), houses sometimes will have access to that water by either "wells" or steps down to the qanat. Basements also can be cooled by qanats used in conjunction with wind towers (figure 9.5) (Roaf 1982). Wealthy homes are usually the ones that have access to qanats (Bonine 1980a). Because very little water is used for domestic purposes, and as long as the water is not contaminated by its use by the household(s), there usually are no complaints by the inhabitants of the settlement to which the qanat flows. In fact, it would be considered immoral—if not illegal under Islamic law—to deny any individuals or households access to water for drinking and domestic purposes.

Where there is sufficient water and enough slope, water-powered grist mills are sometimes built into qanat systems as well (Beazley and Harverson 1982:73–87). Certain public structures are also built or planned within the systems. If a water cistern, public bath, fountain for a mosque, or even a traditional icehouse (Beazley and Harverson 1982:49–56) were present, then clean water would have to be made available from the system. In Khoranaq (northeast of Yazd) there were many separate uses of the water: first for drinking and cooking, then in pools for washing of dishes with sand, then to the public bath, then to small pools for washing clothes (with detergent today), then through three water grist mills for grinding flour,

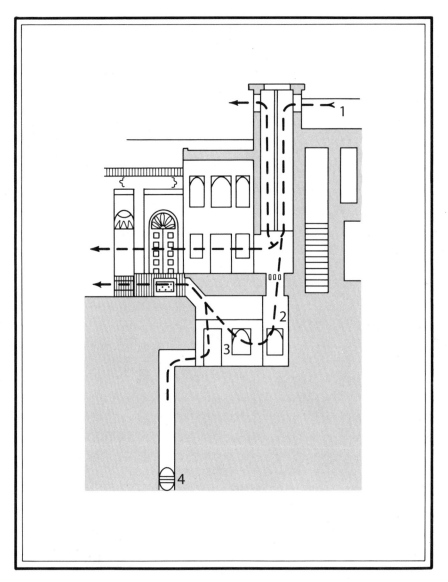

Figure 9.5 Cross-section of an Iranian house, showing how a qanat can cool a basement (adapted from Roaf 1982).

then into pools where straw can be soaked for construction and animals watered, and then finally flowing out to the agricultural fields (Roaf 1989). Most of these uses of qanat water were planned and constructed when the initial system was built.

Qanats also are responsible for "social gradients" within the settlements (English 1966). A social hierarchy is produced related to access to water. Further upslope the water in the system is cleaner and more plentiful, so wealthier and more powerful people, including the landlords, have their houses at the uppermost part of the channel after it emerges from its underground conduit. Especially if there is one main owner, the channel may run through his house (and orchards) before continuing to the rest of the village and the cultivated fields. In linear mountain villages the hierarchy may continue to the end of the housing, where the poorest receive the least and worst water.

Social gradients are also present in cities, where specific neighborhoods or quarters are more or less desirable, depending upon when they receive water. In medieval Nishapur, for instance, the desirability of residential districts was in relation to access to water (Bulliet 1976). In Herat, Afghanistan, the traditional quarters of the old city are even named by the order in which they received water from the Hari Rud, with the last or fourth quarter being the worst residential area of the old city (English 1973). Where several qanats supply a settlement or new qanats are constructed, multiple foci for social gradients are created (see McLachlan 1988). Hence, the morphology of large settlements becomes quite complex, with numerous better or worse neighborhoods being related to the water supply.

Qanats and Wells: The Competition for Water

Compared to the 1950s and 1960s, qanats in Iran have declined substantially in number and, on the average, their flow has decreased. The principal reason for the decline and demise of many qanats has been the construction of modern wells that use diesel pumps. Shallow wells (*chah-i nim-i amigh*), from 5 to 50 meters in depth, and deep tube wells (*chah-i amigh*), from 50 to 150 meters (or even more) in depth, have led to the deterioration of many of these traditional irrigation systems. Basically, too many wells have been dug too close to the mother wells of qanats and too close to one another, and as a consequence there has been a drastic fall in the underground water table in almost all areas of the central plateau.

As the water table falls, the water bearing zone decreases and hence the flow of the qanat is less, creating the same effect as a drought. Finally, as the water table continues to decline, it drops below the mother well and the conduit of the qanat, and the system goes completely dry. Sometimes the channel is extended (i.e., dug farther back into the alluvial fan or plain), but in only a few instances has this operation been successful. Wells have caused drops of 10 to 20 meters in the water table in only a few years, and in some cases over 100 meters within one or two decades. In such situations, qanats, some of which had operated effectively and efficiently for centuries, have become useless dry channels.

If a village's qanat dries up, the village will die, so the settlement must now dig a well to survive. In some cases such a well will in turn affect the qanats that flow to other villages downslope. To survive, more and more wells must be dug and the many wells begin to compete with each other. A vicious cycle is created in which an entire system or valley of qanats will soon be destroyed. Ehlers (1989) provides an excellent example of the introduction of wells and their impact on the qanats of an entire valley in Assadabad, a basin of the central Zagros. The valley covers about 660 square kilometers and has 130 villages, 83 of which can be termed mountain villages and 48 villages of the plain. In 1976 these villages supported about 65,000 persons and cultivated 35,000 hectares annually. About half of the villages (and population) were originally watered by qanats, with the remainder being supplied by springs and small streams. Nearly all of the villages in the plain, which were the larger settlements, had qanats.

The first shallow well was introduced into the valley as early as 1945, but by 1962 the number of wells had increased to thirty-nine. Then the number increased dramatically—to 439 in 1969 and to 1,386 in 1979. The depth of these wells in most cases ranged from only 8 to 12 meters. But by 1979 at least eighteen deep tube wells had been constructed; these wells ranged from 70 to 118 meters in depth. They had a much greater discharge than the shallow wells and affected the water table much more rapidly.

Ehlers (1989) also notes that a deep well costs on the average 175,200 tomans ($23,000)—with annual management and maintenance costs of 20,000 tomans ($2,600). On the other hand, a shallow well costs only about 22,000 tomans ($2,900) to construct and considerably less to maintain. These figures emphasize that, similar to qanats, considerable capital was needed for a deep well, although some peasants (or several together) could more easily manage to finance the digging of a shallow well.

All of the qanats of the Assadabad Basin were destroyed by the late 1970s. The water table had dropped so drastically that a number of the shallow wells had also dried up. Even more disturbing is the fact that the agricultural system declined as a result of the new dependence on wells and the subsequent overuse of the aquifer by the wells. Also, with the disappearance of the qanats the social order changed. The new and more productive deep wells, which required considerable capital, were financed and owned by large landlords. They now had their own substantial supply of water in which the common villagers did not have a share (as they had had under the qanat water arrangements). These landlords irrigated their own lands and changed to mechanized agriculture, the latter exempting their land for redistribution under the terms of the land reform program of the early 1960s. More peasants were now without any cultivation rights, and many were in debt as they tried to finance the digging and maintenance of the shallow wells, which in some cases went dry anyway.

In summary, in only two decades, over a thousand wells were dug in the Assadabad Valley, causing a drastic decline in the level of the water table and ruining all of the qanats. It was a major ecological and socioeconomic disaster—and one that, unfortunately, was repeated in many other valleys and regions of Iran. Since the early 1980s, on the other hand, the Islamic Republic has focused its agricultural policies at improving local village agriculture in its goal for self-sufficiency in agriculture (McLachlan 1988; Schirazi 1993), so perhaps the decline of some of the systems has been reversed (hence, the statistics for the increase in qanats in the 1980s mentioned earlier).

Qanats as Viable Local Systems

Qanats in Iran are the legacy of a traditional technology that was developed over 2,000 years ago (English 1968), and they are an integral part of traditional Iranian rural society and culture. It is a system that has enabled human societies to occupy and survive in extremely marginal arid environments. It also represents a system that is in equilibrium with the environment. Yet, qanats often have been considered anachronistic in today's modern technological world. Planners, agricultural engineers, consultants, and development agencies often have viewed these underground water systems as not conducive to modern, capitalist commercial agriculture. In fact, the decline of the qanat in Iran in the 1960s and 1970s was facilitated by the Pahlavi government's agricultural (and land reform) policies, which promoted mechanization of agriculture and development of large-scale agricultural projects for commercial export crops (Hooglund 1982; Goodell 1986; McLachlan 1988).

The belief often has been that "qanats are relics of an age when only small-scale power sources were available and when making use of gravity to deliver irrigation water was highly beneficial. They played a vital role in sustaining agriculture in semi-arid and arid regions, but in the future it seems inevitable that they will become less and less important" (Beaumont 1989: 31). But is it inevitable that qanats will decline, become less important, or disappear completely?

The advantages and disadvantages of these systems must be understood and evaluated. Disadvantages are that

1. the systems are expensive and time consuming to build,
2. water flows, even when not needed,
3. a group of specialists (muqanni) are needed to keep the qanats in repair,
4. the systems are subject to droughts and seasonal changes,
5. water is lost by seepage in the channels,
6. planning for agricultural development is difficult because of the unreliable water supply,
7. qanats are limited to only certain areas of sufficient aquifers, slope, and so on,
8. they can be dangerous for the workers, and
9. large storms can damage them, requiring difficult, expensive repairs.

Advantages of qanat systems include

1. a fresher water source than is available at lower elevations,
2. gravity water flow that needs no pumps or other mechanical equipment,
3. mostly underground water flows, which minimizes evaporation loss,
4. low maintenance costs (once built), with only occasional cleaning needed,
5. appropriate use of the aquifer that does not change the availability of water,

6. small size, which makes them compatible with small villages and local management,
7. making water available for small-scale farming, thus minimizing the need for large schemes that overuse water and damage the aquifer and ecology for the short term,
8. promoting the continued viability of farming communities by being an integral part of the rural social system,
9. contributing to a sustainable irrigation system for entire valleys by allowing water seepage to be used by other qanats farther downslope, and
10. contributing to the growth of vegetation along channels and within villages when channels are at the surface (a situation that would not occur with efficient, closed pipes).

Perhaps a fresh look and a new evaluation of qanat systems are needed. If combined with modern technology these systems might continue to provide a viable source of water for the future. Advancements and improvements might include modern surveying instruments, boring machinery, the use of fans for ventilation for workers, the use of better lining for a safer channel, the use of concrete pipes to convey water in certain areas to reduce seepage, and the construction of reservoirs to save the water in the winter. Also, recent attempts at artificial recharge of aquifers of qanats have enabled the systems to continue or even increase in flow (Behnia 1993). Even with technological changes, qanats can still remain manageable at the local level.

Certainly, the advantages and disadvantages of qanats must be considered and evaluated in order to determine whether or not these systems can be viable in the modern world. Yet it is apparent that modern pump wells, although more "efficient" than qanats, have led to major ecological and social problems in Iran. As in many dry regions of the world, the overpumping of aquifers and even depletion in some cases have led to the eventual decline of agriculture and to the deterioration of the environment and increased desertification. In the case of Iran it also has promoted the dismantling of rural society, changing the costs and access to water, and even the socioeconomic interrelationships of villagers.

Although there are some problems and disadvantages to qanats, there also are many positive benefits of these local, farmer-managed irrigation systems. Their preservation and continued use need to be incorporated into any future irrigation and agricultural planning in Iran. Such systems, which can be properly managed at the local level and provide a livelihood for considerable numbers of people, should not be deemed anachronistic and unsuitable for the modern world. They have proven their sustainability in the extremely marginal arid environment of the Iranian Plateau.

References Cited

Ajami, Amir I.
    1993   Cooperatives. *Encyclopaedia Iranica* 6: 253–58.
Alberts, Robert Charles
    1963   *Social Structure and Culture Change in an Iranian Village.* Ph.D. dissertation, University of Wisconsin, Madison.
Beaumont, Peter
    1971   Qanat Systems in Iran. *Bulletin of the International Association of Scientific Hydrology* 16: 39–50.
    1973   A Traditional Method of Groundwater Extraction in the Middle East. *Ground Water* 11:23–30.
    1989   The Qanat: A Means of Water Provision from Groundwater Sources. In *Qanat, Kariz and Khattara: Traditional Water Systems in the Middle East and North Africa,* ed. P. Beaumont, M. Bonine, and K. McLachlan, 13–31. Cambridgeshire: MENAS Press.
Beaumont, Peter, Michael Bonine, and Keith McLachlan (editors)
    1989   *Qanat, Kariz and Khattara: Traditional Water Systems in the Middle East and North Africa.* Cambridgeshire: MENAS Press.
Beazley, Elisabeth, and Michael Harverson
    1982   *Living with the Desert: Working Buildings of the Iranian Plateau.* Warminster, Wilts, England: Aris and Phillips.
Behnia, Abdul Karim
    1993   Artificial Recharge of Karez. In *The Proceedings of [the] International Conference on Karez Irrigation,* ed. International Conference on Karez Irrigation, 229–31 (English Section). Urumqi, China: Xinjiang People's Publishing House, Educational and Cultural Press.
    1980s   *Qanatsazi va Qanatdari* [Qanat Construction and Qanat Maintenance]. Tehran: University of Tehran, Central Publishing [House].
Bobek, Hans
    1974   Zum Konzept des Rentenkapitalismus. *Tijdschrift voor economishe en sociale geografie* 68:73–78.
    1979   Rentenkapitalismus und Entwicklung in Iran. In *Interdisziplinare Iran-Forshung.* Beihefte zum TAVO, Reihe B (Geisteswissenschaften), No. 40., ed. Gunther Schweizer, 113–24. Wiesbaden: Dr. Ludwig Reicher Verlag.

Bonine, Michael E.

1979   The Morphogenesis of Iranian Cities. *Annals of the Association of American Geographers* 69:208–24.

1980a   Aridity and Structure: Adaptations of Indigenous Housing in Central Iran. In *Desert Housing: Balancing Experience and Technology for Dwelling in Hot Arid Zones*, eds. K. N. Clark and P. P. Paylore, 193–219. Tucson: Office of Arid Lands Studies, University of Arizona.

1980b   *Yazd and Its Hinterland: A Central Place System of Dominance in the Central Iranian Plateau.* Marburger Geographische Schriften No. 83. Marburg: Geographischen Institut der Universitat Marburg.

1982   From Qanat to Kort: Traditional Irrigation Terminology and Practices in Central Iran. *Iran* 20:145–59.

1989   Qanats, Field Systems, and Morphology: Rectangularity on the Iranian Plateau. In *Qanat, Kariz and Khattara: Traditional Water Systems in the Middle East and North Africa*, ed. P. Beaumont, M. Bonine, and K. McLachlan, 35–57. Cambridgeshire: MENAS Press.

Bulliet, Richard W.

1976   Medieval Nishapur: A Topographic and Demographic Reconstruction. *Studia Iranica* 5:67–89.

Ehlers, Eckart

1978   Rentenkapitalismus und Stadtentwicklung im islamischen Orient: Beispiel Iran. *Erdkunde* 32:124–42.

1980   *Iran: Grundzuge einer Geographischen Landeskunde.* Wissenschaftliche Landerkunden, Vol. 18. Darmstadt: Wissenschaftliche Buchgesellschaft.

1984   Westliches Iran: Landnutzung. *Türbinger Atlas des Verderen Orients (TAVO)*, Map AX3.

1989   Qanats and Pumped Wells—The Case of Assad'abad, Hamadan. In *Qanat, Kariz and Khattara: Traditional Water Systems in the Middle East and North Africa*, ed. P. Beaumont, M. Bonine, and K. McLachlan, 89–112. Cambridgeshire: MENAS Press.

English, Paul Ward

1966   *City and Village in Iran: Settlement and Economy in the Kirman Basin.* Madison: University of Wisconsin Press.

1968   The Origin and Spread of Qanats in the Old World. *Proceedings of the American Philosophical Society* 112:170–81.

1973   The Traditional City of Herat, Afghanistan. In *From Madina to Metropolis*, ed. L. C. Brown, 73–90. Princeton: Darwin Press.

Ghayoor, Hassan Ali

1990   Qanat: A Reconsideration of the Iranian Irrigation System. Unpublished Paper, Department of Hydrology and Water Resources, University of Arizona, Tucson.

Goblot, Henri

1979   *Les Qanats: Une technique d'acquisition de l'eau.* Ecole des Hautes Etudes en Sciences Sociales, Industrie et Artisanat No. 9. Paris: Mouton.

Goodell, Grace E.

1986   *The Elementary Structures of Political Life: Rural Development in Pahlevi Iran.* New York: Oxford University Press.

Hartl, Martin

1979   *Das Najafabadtal: Geographische Untersuchung einer Kanatlandschaft im Zagrosgebirge.* Geographische Schriften No. 12. Regensburg: Institut fur Geographie an der Universitat Regensburg.

1989   Qanats in the Najafabad Valley. In *Qanat, Kariz and Khattara: Traditional Water Systems in the Middle East and North Africa,* ed. P. Beaumont, M. Bonine and K. McLachlan, 119–35. Cambridgeshire: MENAS Press.

Hooglund, Eric J.

1981   Rural Socialeconomic Organization in Transition: The Case of Iran's Bonehs. In *Modern Iran: The Dialectics of Continuity and Change,* ed. M. E. Bonine and N. R. Keddie, 191–207, 419–22. Albany: State University of New York Press.

1982   *Land and Revolution in Iran, 1960–1980.* Austin: University of Texas Press.

Kielstra, Nicolaas O.

1975   *Ecology and Community in Iran: A Comparative Study of the Relations between Ecological Conditions, the Economic System, Village Politics and the Moral Value System in Two Iranian Villages.* Amsterdam: University of Amsterdam Printer.

1989   The Reorganisation of a Qanat Irrigation System near Estehbanat, Fars. In *Qanat, Kariz and Khattara: Traditional Water Systems in the Middle East and North Africa,* ed. P. Beaumont, M. Bonine, and K. McLachlan, 136–49. Cambridgeshire: MENAS Press.

Lambton, A.K.S.

1953   *Landlord and Peasant in Persia.* London: Oxford University Press.

1989   The Origin, Diffusion and Functioning of the Qanat. In *Qanat, Kariz and Khattara: Traditional Water Systems in the Middle East and North Africa,* ed. P. Beaumont, M. Bonine, and K. McLachlan, 5–12. Cambridgeshire: MENAS Press.

McLachlan, Keith

1988   *The Neglected Garden: The Politics and Ecology of Agriculture in Iran.* London: I.B. Tauris.

Roaf, Susan

1982   Wind-Catchers. In *Living with the Desert,* by Elisabeth Beazley and Michael Harverson, 57–72. Warminster, Wilts, England: Aris and Phillips.

1989   Settlement Form and Qanat Routes in the Yazd Province. In *Qanat, Kariz and Khattara: Traditional Water Systems in the Middle East and North Africa,* ed. P. Beaumont, M. Bonine, and K. McLachlan, 59–60. Cambridgeshire: MENAS Press.

Sadr, Kazem

1357/1978–1979   Ahamiat-i Konuni-yi Qanat [The Present Significance of Qanats]. Paper presented at the First Symposium of Iranian Agricultural Policy, University of Shiraz.

Safinezhad, Javad

    1977   The Climate of Iran and the Emergence of Traditional Collective Production Systems. Unpublished paper in possession of the author.

    1356/1977–1978   Qanat dar Iran [The Qanat in Iran.] *Journal of Daneshkadeh* 8.

    1359/1981–1982   *Nezamha-yi Abiari-yi Sonnati dar Iran* [Traditional Irrigation Systems in Iran]. Tehran: Institute for Social Studies and Research.

Schirazi, Asghar

    1993   *Islamic Development Policy: The Agrarian Question in Iran.* Boulder, Colo.: Lynne Rienner Publishers.

Spooner, Brian

    1974   Irrigation and Society: The Iranian Plateau. In *Irrigation's Impact on Society,* ed. T. E. Downing and McG. Gibson, 43–57. Tucson: University of Arizona Press.

Wulff, H.E.

    1968   The Qanats of Iran. *Scientific American* 218(4):94–105.

10

The Utility of Tradition in
Sri Lankan Bureaucratic Irrigation
The Case of the Kirindi Oya Project

Pamela C. Stanbury

Irrigated agriculture has a long history in Sri Lanka. The ancient hydraulic civilization dating to the second century B.C. is a testament to the country's achievements in irrigation engineering and social organization for irrigation management (Murphey 1957; Gunawardna 1971). Although many of the ancient irrigation systems were abandoned during the thirteenth century A.D., in recent years the government of Sri Lanka has rehabilitated them and settled them with people from the overcrowded coastal zones. These "new" settlements have come to be known as the "colonization schemes." A total of 263,636 hectares of newly irrigated land has been settled in the last fifty years (Stanbury 1989:3). Today, public spending on irrigation constitutes one of the largest parts of the Sri Lankan government's budget, underscoring the importance of irrigation to the economy.

Despite the government's investment in irrigation, the new colonization schemes are characterized by numerous management problems. To many critics, the fault lies in the lack of farmer involvement in managing the systems and the farmers' strong dependence on the government bureaucracy. Efforts to remedy management problems have focused on enhancing water-user participation in the operation and maintenance of irrigation systems. And the key has been to incorporate many social and institutional features that are thought to have existed — and worked effectively — in traditional village irrigation systems of the past. The result has been the use of idealized images of the past to reengineer the social fabric of irrigation management in Sri Lanka.

These efforts call into question the utility of applying principles of traditional village irrigation to the bureaucratically managed colonization schemes of today. Can traditional irrigation management institutions be called upon to effectively improve the performance of large, complex systems? Are traditional irrigation communities so different from the large, state-run colonization schemes that any effort to apply features of the latter to the former is futile? Skeptics often argue that modern bureaucratic systems lack the social characteristics of traditional systems, particularly those concerned with control over water. In this view, the social and economic fabric of irrigation communities has changed so drastically that any attempt to replicate traditional irrigation management practices will not work (Hunt 1985; Abeywickrema 1988; Abeyratne 1986; J. Perera 1988).

This chapter posits a more optimistic view about the utility of applying traditional village irrigation management principles—whether idealized or not—to today's colonization schemes. Clearly, there are great differences between the two systems, yet perceptions of traditional irrigation management have stimulated government attempts to pursue a more participatory approach to water management and involve water users more in irrigation decisions. From this perspective, the lessons from traditional systems may be providing the impetus for a bureaucratic reorientation on the part of government staff, greater communication between irrigation officials and water users, and greater cooperation among users.

I begin this chapter with an overview of modern irrigation in Sri Lanka and the growth of the colonization schemes. I then describe the elements of traditional irrigation management and show how these have been transposed on a new colonization scheme in southern Sri Lanka, known as the Kirindi Oya Irrigation and Settlement Project (KOISP) (figure 10.1). I conclude with an assessment of the value of applying traditional village irrigation management principles to systems characterized by enduring bureaucratic control.

Modern Irrigation In Sri Lanka

The Sri Lankan economy is predominantly agricultural. Irrigated rice is one of the principal crops grown and provides the staple food for most households. Other major crops are tea and coconuts. Geographically, the island is divided into two distinct zones: a wet zone in the coastal southwest region and a dry zone occupying approximately 60 percent of the land area. Rainfall in the Dry Zone is relatively limited and is concentrated in two periods from October to December and April to May. The topography and climate of the Dry Zone permit irrigated rice cultivation. High constant temperatures, gentle relief, concentrated rainfall and a succession of small shallow stream valleys combine to promote irrigation based on water captured in small tanks or reservoirs (De Silva 1981:28). In areas where the topography permits, larger tanks can support greater

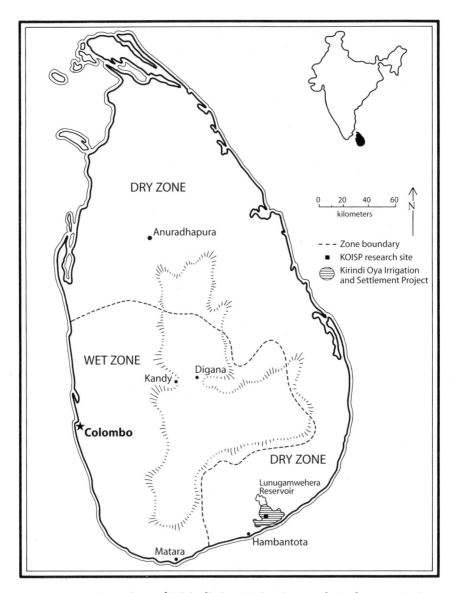

Figure 10.1 Location of Kirindi Oya Irrigation and Settlement Project (KOISP), Sri Lanka.

population concentrations. Irrigation has played a vital role in the Dry Zone throughout history. The remains of the ancient hydraulic civilizations dating from the second century B.C. to the thirteenth century A.D. attest to the area's potential. Today, the total land area under irrigation is 500,000 hectares, including small village tanks less than 80 hectares, larger colonization schemes, and complex multipurpose systems covering 40,000 hectares and more.

To manage and service these complex and multiple systems, an irrigation bureaucracy has grown increasingly complex. The small village tank schemes (known as minor schemes) are, in principle, managed by the local community, but the Department of Agrarian Services provides assistance as required. All other types of systems are controlled and managed by separate agencies within the government bureaucracy. The colonization schemes, which cover a total of approximately 300,000 hectares of land, are managed by the Irrigation Department and large multipurpose schemes are managed by specialized departments, modeled after the Tennessee Valley Authority. The well-known Mahaweli Project is managed by the Mahaweli Authority of Sri Lanka (MASL) and covers approximately 460,000 hectares (Perera 1986:1).

Bureaucratic Origins

The irrigation bureaucracy has its origins in the nineteenth century, when British colonial administrators, intent on increasing agricultural revenues, began to focus on the Dry Zone. Long abandoned since the thirteenth century, the visible remains of the ancient irrigation works inspired colonial administrators to recapture the past glory and create a newly flourishing agricultural sector. Attempts to rehabilitate the abandoned tanks began, accompanied by efforts to resettle people from the more densely populated coastal areas. In 1900, the Irrigation Department was established to manage these new colonization schemes.

In the 1930s, severe food shortages throughout the country led government officials to seek ways to promote a rapid increase in agricultural production. The result was full-scale colonization in the Dry Zone, a process that has been well documented by B. H. Farmer (1957). Since the 1930s, colonization has become an increasingly politicized issue. For many of the Sinhalese majority, it has become a symbol of a return to the heartland of the ancient Sinhalese irrigation civilization (Moore 1985:45).

The Dependency Syndrome

Initiating colonization schemes in Sri Lanka involved a good deal of government control and social engineering. Government officials provided incentives to move to what, in many cases, were less than desirable locations. The settlements were often remote, lacked amenities, and were far removed from family and existing social networks. In order to attract would-be settlers, the government created the basic infrastructure, including land for farming, homesteads and irrigation water. It financed the irrigation works, conducted land leveling and initiated the jungle clearing. Furthermore, as part of its welfare attitude towards the peasantry, the government selected settlers who were landless, gave them equal amounts of land in new communities and a package of inputs to start a new life. Settlers were not required to pay for irrigation water, and all financial responsibility was assumed by the state.

The high level of government involvement in the colonization schemes has been attributed to physical and institutional factors associated with the large scale of these schemes. Physical factors include the demands of large and complex irrigation distribution networks, the large number of small-scale farmers involved, and the great distance between the main system and the beneficiaries, so that the main systems could not be easily managed by the communities themselves. Institutional factors include the government's social welfare orientation, the heterogeneous nature of the settlers, the inability of large groups to organize cultivation calendars themselves, and the insensitivity of the irrigation bureaucracy to farmers (Abeywickrema 1988:23).

Not surprisingly, these schemes came to be characterized by a great deal of settler dependency on the state, and government officials played a paternalistic role with settlers. The level of dependency was demonstrated in an extreme form when officials even chose marriage partners for the settlers (Ellman et al. 1976:8). Settlers, on their part, tended to view the schemes as belonging to the government, not themselves. The result was that settlers have felt no sense of ownership of the irrigation channels and have felt no reason or need to participate in irrigation maintenance or rehabilitation.

During the 1960s the poor returns on investments and the gross inefficiencies in management of the colonization schemes became increasingly clear. Systems fell in disrepair, water was wasted, and water distribution problems became common. Many critics began to argue that the root cause of the problem was the absence of effective local level organizations and leadership within the irrigation communities. They criticized the government for its overly domineering role that effectively undermined any community management and engendered dependency on outside forces (Merrey and Somaratne 1988). Since the 1970s, efforts to redress these concerns have focused on promoting greater participation in irrigation operation and maintenance by the colonists themselves. The logic is that if farmers participate more in decision making, they will take an interest in making needed repairs, invest in routine maintenance and allocate water more efficiently. Furthermore, in doing this, farmers will reduce government expenditures and optimize agricultural production and incomes.

Several experiments involve farmers more in irrigation decisions at the tertiary and secondary level of the canal systems. These experiments involve the promotion of water-user associations, which are designed to increase farmer involvement in planning and design, water distribution, and maintenance and rehabilitation. The water-user associations are formal organizations of water users, with elected representatives, an active membership and regular meetings. Perhaps the most important experiment in this regard has been the Gal Oya Project, which was established to rehabilitate one of the old tanks and to bring new land under cultivation. At Gal Oya, efforts were made to foster water user groups using "catalysts." The catalysts were local field officers, trained in participatory management, who worked closely with water users to help them develop the water-user associations along field channel and distributary channel lines (Perera 1988; Uphoff 1986).

The experiment at Gal Oya and subsequent experiments in other major irrigation schemes suggest that farmers will respond to opportunities to take greater responsibility for system operation and maintenance. These experiences also suggest that the presence of legitimate and effective farmer associations can lead to improved cooperation among farmers and improved cooperation and communication between farmers and agency officials.

Tradition in Modern Irrigation Systems

Efforts to foster farmer participation in the colonization schemes today, such as the Gal Oya experiment, have in part drawn on the experience in other Asian countries (see Uphoff 1986; ISPAN 1994). Yet they have also borrowed specific features of Sri Lanka traditional village irrigation systems and have applied them to the modern context. These features arise from the locally managed village tanks. Historically, village tanks served local communities under the principle of one tank–one community. A village tank was created by damming up a natural stream and building a long earthen wall to hold the water. At various periods of decline in the central government, many of the small village tank systems endured while the larger, centrally controlled ones were more prone to collapse (De Silva 1981).

Modern perceptions of the traditional village irrigation system are generally as follows:

1. The community depends heavily on the irrigation system for agriculture as well as domestic needs, inducing a sense of community responsibility for the whole system.
2. The village community is relatively homogenous.
3. Because of its small size, the farmers can control, manage, and have an intimate knowledge of the whole system.
4. Traditional systems include the *kanna* (cultivation) meeting, which provides for farmer participation in the planning of the cultivation calendar and the enforcement of cultivation decisions.
5. Also included is the *bethma,* a practice of sharing irrigable land during periods of drought and limited water supply.
6. The village organization and the *vel vidane* (irrigation headman) system ensure proper administration and equitable distribution of water.
7. Labor and payments in kind (*rajakariya*) are contributed to ensure proper maintenance of irrigation facilities.

This characterization conforms to Edmund Leach's classic description of village irrigation in Pul Eliya, and indeed, one wonders if Leach himself was not responsible for perpetuating beliefs about traditional irrigation management. Leach (1961) described a community oriented around the village irrigation tank. According to Leach, the group was relatively homogeneous and community members were all stakeholders in the irrigation system. Kinship, as he described it, was not so much a function of descent from a common ancestor as it was one of location—that is, location around the tank. Within the community of Pul Eliya, a number of institutions ensured the operation and maintenance of the system. A public village meeting, the kanna meeting, provided a forum for collective decisions about water distribution and cropping patterns. The kanna meeting exists today and is one social institution that has been formally codified in irrigation legislation; it is required of all irrigation systems, both small village schemes and larger multipurpose schemes.

Leach also described the responsibilities for water distribution and management in Pul Eliya. A locally elected leader, the vel vidane (irrigation headman), was responsible for operating the sluice and distributing water. The distribution system was calculated by tradition, not something that the vel vidane worked out for himself (Leach 1961:164). Leach (1961: 165) notes, "although the present generation of Pul Eliya villagers are not at all clear about the inner logic of it all they are keenly aware that the numerical formulae handed down from ancient times are very important." The concept of the vel vidane has been the subject of much debate, in terms of its historical roots (Hunt and Hunt 1976:394) and its utility as an institution today (Alwis 1988). At some points in history, the position of the vel vidane was legally sanctioned, while during other times the sanction was removed. Today, efforts are being made to ensure that the legal definition of this position fits within the context of the major colonization schemes and to ensure that it serves to represent the water users' interests.

Bethma, the system of sharing irrigated land during times of water scarcity, was practiced in Pul Eliya to ensure equitable distribution of water, according to Leach. This practice was enforced through customary arrangements, though not codified in law. In recent major irrigation schemes, the government has tried to revive the practice of bethma in efforts to conserve water and distribute it more equitably (see Ekanayake and Groenfeldt 1987).

The bureaucratically run systems of today are in many ways different from the traditional village tank schemes. Yet, the government has introduced elements to foster greater community participation, improve operation and maintenance, and reduce water-user dependency on government agencies. How well the principles of traditional village irrigation management fit the context of modern day bureaucratically run systems remains a key question for irrigation planners and policy makers. The recent Kirindi Oya Irrigation and Settlement Project (KOISP) offers an opportunity to see how well the traditional social institutions for irrigation management fit in the modern context.

The Kirindi Oya Irrigation and Settlement Project

The Kirindi Oya Irrigation and Settlement Project (KOISP) is one of the Sri Lankan government's most recent efforts to bring new land under cultivation through irrigation development and colonization. The project is situated in southern Hambantota District, approximately 260 kilometers from the capital city of Colombo along the coastal road (figure 10.1). The project involves the development of water resources of the Kirindi Oya (*oya* means river in Sinhalese) and adjacent areas. It is designed to provide irrigation facilities, social services, and infrastructure to a newly settled population of approximately 8,000 families. In addition, the project involves augmentation of the water supply of six existing tanks and rehabilitation of irrigation structures in an adjoining settled area of 4,585 hectares.

KOISP lies in the Dry Zone, with a mean temperature of 26° Centigrade, and annual rainfall of less than 1,230 millimeters. Seventy-five percent of rain falls during the *maha,* or rainy season, from October to March. Rainfall in the area is so seasonal and erratic that cultivation of rice is untenable without irrigation. Thus, the introduction of irrigation has made permanent agriculture possible during the maha. If enough water remains in the reservoir, irrigated crops may also be cultivated during the dry season from April to August, known as *yala.*

Initial planning for KOISP began in the 1950s as part of a more comprehensive government scheme to develop the water resources of eight major river basins. Financial and political setbacks delayed the project until 1981. Phase I was completed in 1988 with settlement of approximately 4,000 families and allocation of an equal number of hectares of irrigated land. The project has received substantial financial assistance from the Asian Development Bank and the German Kreditanstalt für Wiederbau (KFW). In large measure, KOISP is typical of recent irrigation development in Sri Lanka—large in scale and dependent on foreign involvement in financing and planning.

The project is located in an area with archaeological remains of irrigation works, an indication of the areas' irrigation potential. Much of the area was abandoned, and in 1881 a British chronicler described it as a "desolate wilderness" (Farmer 1957:13). After six centuries of disrepair, several of the ancient irrigation tanks were restored by the British in the late nineteenth century, and a diversion scheme was developed to tap water from the Kirindi Oya. Village tank irrigation in the area is described by Harriss (1977). Under the KOISP, the water supply for these tanks has been augmented with supplies from the river. In addition, many of the old irrigation structures have been rehabilitated. Thus, the old tanks are now linked to a larger irrigation network to achieve a more secure and dependable water supply.

A complex bureaucracy developed to manage and support the settlement and irrigation activities in KOISP. The Irrigation Department, Land Commissioner's Department, and numerous other agencies were all involved in the construction of irrigation systems, settlement and infrastructure development and provision of services to farmers. In addition to these agencies, a newly created Irrigation Management Division was given responsibility for fostering community participation in irrigation management tasks. This division was established at the central government level in 1984, with the expressed mandate of fostering community participation and coordination among government agencies.

Community Management in KOISP

Because it was one of the last irrigated settlement schemes to be developed, KOISP profited from lessons learned during previous decades of experimentation with different approaches to colonization. Drawing on the failures of the past, the KOISP managers made efforts to engage the community of new settlers to participate in managing the irrigation system. Although some of the techniques drew upon experiences elsewhere in Asia (most notably, the Philippines), many of the techniques also have a decidedly Sri Lankan cultural foundation and are rooted in Sri Lankan village community traditions. These techniques have had to be modified to meet the demands of a large bureaucratically controlled system but still show promise for improving the operation and maintenance of KOISP.

The One-Tank-One-Village Principle

The Irrigation Department and Land Commissioner's Department coordinated efforts to create hydrologically based communities so residence and water sharing would be coterminous. This attempts to replicate the one-tank-one-village principle described for traditional village irrigation by Leach (1961). Although KOISP is a much larger system than traditional village tank systems, it has been divided into smaller community units. Thus, the two main canals (left and right bank main canals) are divided into tracts with approximately four residential villages and twelve distributaries per tract. Each village corresponds to approximately three distributary channels. Finally, a number of field channel turnout units are located along each of the distributaries. Each turnout includes approximately five to fifteen individual farm allotments.

The selection of settlers and allocation of land to them was designed to foster cooperation in managing shared water resources. New settlers were each allocated one hectare of irrigated land for cultivation of rice and a smaller half hectare of highland for a homestead in the adjacent village. Settlers who came from the same villages originally were allocated land near one another. Furthermore, planners attempted to allocate homesteads to groups of farmers who had irrigated allotments on the same turnout of the canal system.

The significant point is the government's attempt to create a structure that would foster community participation in the operation and maintenance of the irrigation system. It marked a significant change from earlier efforts to get the engineering design right, but at the expense of community organization. In KOISP, planners tried to avoid this by fostering hydrologically based communities within the larger project.

Creating Water-User Associations

With a community structure in place, the next step was to foster active participation in irrigation management by the new communities of settlers. The Irrigation Management Division (IMD), a separate division in the large irrigation bureaucracy, undertook this task. A project manager from IMD was assisted by a number of Association Organizers (AOs), who resided in the villages and worked closely with community members to form water user associations. This catalyst approach was developed in Gal Oya and had begun to spread around the country to other schemes.

In KOISP, each water-user association conformed to a hydrological unit. The structure was hierarchical and corresponded to the turnout and distributary levels of the system. Each field channel comprised a group with an elected field channel representative. Each field channel representative was then a member of a distributary organization that had an elected distributary channel representative. In turn, each distributary channel representative formed part of a village level association (figure 10.2). Water distribution in the turnout area was handled by the association members. The turnout representatives were responsible for coordinating the distri-

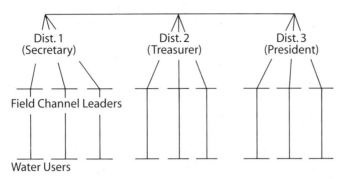

Figure 10.2 Organization of water-user associations in the Kirindi Oya Irrigation and Settlement Project.

bution of water within the field channel. At the time of field research, it was unclear whether water users would eventually take responsibility for distributary channel water distribution.

The farmer representative role bears some similarities to the role of the vel vidane described in traditional village tank systems. The vel vidane played a key role as irrigation headman, elected by consensus of the community members, but ultimately under the control of the irrigation bureaucracy. In KOISP the farmer representatives were independent of the government. As elected leaders they could voice concerns and problems to the irrigation officials but were not directly under the irrigation bureaucracy's control.

The Kanna Meetings
The planning procedures for the cultivation season were developed at the kanna meeting in KOISP. But, in a departure from the village tank schemes, the planning procedures in KOISP required collaboration between government officials from the irrigation department and the water users. The dates were officially fixed at the meeting and decisions were made about the extent of the area to be cultivated. The kanna meeting was a time when project officials and farmers could reach consensus. Pre-kanna meetings were held by the irrigation tract members, to enable all farmers in each tract to participate and to get farmers' approval. The pre-kanna meetings allowed farmers to express their grievances and resolved disputes prior to the kanna meeting itself (Merrey and Somaratn 1988).

Shramadana

In KOISP, each turnout group was expected to participate in desilting channels, making minor repairs, and contributing to maintaining drainage areas through labor contributions and work groups. Distributary maintenance was also to be carried out by the associations, but through paid contracts from the Irrigation Department. This form of traditional labor contribution met with limited success in KOISP. The Irrigation Department complained that it was too difficult to get farmers to do the work on time. Although farmers did participate in some work on the field channels, many felt that the problems of alignment, siltation, and drainage were due to the design and construction problems and pointed the finger at the Irrigation Department for deficiencies. There was the sense that the system belonged to the bureaucracy, not to them.

Bethma

Although the practice of bethma did not exist in KOISP, planners discussed implementing some form of land sharing as new calculations about the amount of water available in the reservoir became available. Once the second phase of the project was completed, it was anticipated that some form of bethma would be necessary to make up for the inadequate supplies of water. Whether bethma is a feasible solution in a large system like KOISP remains to be seen.

The Value of Tradition in KOISP

KOISP, like other recent colonization schemes, differs from the image of traditional village tank systems in that water users do not have control over the entire system and must deal with an external government bureaucracy. A number of technical factors make it difficult for community members in the larger colonization schemes to undertake all responsibility for water delivery. First, the maintenance of the headworks requires the skills of a trained professional. Management of the main canal system requires the services of paid officers, and if the main system does not work properly, farmers cannot manage water below the field channel level. Furthermore, settler involvement in management decisions is hampered by the fact that they are a heterogeneous group and often lack the in-depth knowledge of the local area and knowledge of how the irrigation system works.

Given these factors, a select number of features of traditional village irrigation management have been adapted and incorporated into KOISP. The hydrologically based settlement planning, the kanna meeting, water-user representation, and potential labor contributions by water users all set the stage for more effective cooperation and participation in managing a common water resource at the second and tertiary level of the irrigation system. Farmer representatives are a modification of the vel vidane position, modified so that the position truly represents the water users. This is important in a system where external bureaucratic control is persistent. Although at the time of this field study the representatives still had difficulty communicating their needs to agency officials, the institutionalized nature of the vel vidane position helped water users voice their problems and concerns so the system could work more effectively for them.

KOISP was still in an early stage of development during the period of field research, and the newly formed water user associations were not yet established. Some problems of cooperation for maintenance and water distribution were evident, but were largely attributed to the fact that the system was new and construction was still ongoing. The future remains to be seen, but KOISP shows that the principles of irrigation management attributed to traditional systems may be applied usefully to modern bureaucratic systems.

Perhaps most important, the application of traditional irrigation management principles in KOISP has contributed to a bureaucratic reorientation within the irrigation agencies. Although the farmers' ability to effect change was limited because they only represented parts of the system, they were able to draw attention to problems in ways that would have been impossible without the water user associations. Agency staff were beginning to recognize the need to work with water users and to listen to their problems and demands, a step in the direction of a more participatory and democratic management structure.

Conclusions

Like many Asian countries, bureaucratic control dominates much of Sri Lanka's modern irrigation practices. Yet, recent efforts have been made to reduce farmer dependency on state agencies and involve them more in managing the irrigation systems. Many of the ingredients of participatory management are found in the customary practices described for traditional village irrigation systems. KOISP and systems like it clearly differ from the village tank systems in their size, issues of control over water, planning, and design. These factors pose new challenges in terms of system operation and maintenance.

This chapter has attempted to show some linkages between traditional and modern irrigation principles and practices. It takes the position that these features have been important in the bureaucratic reorientation of the government agencies in Sri Lanka. Customary practices have provided a model for changing behavior of agency staff and for changing farmers' attitudes towards the agencies and the irrigation system itself. If it appears an overly optimistic scenario, it is because the author sees a role for traditional irrigation management in irrigation development today. Whether real or idealized, the images of participation, of community organization and of community responsibility provide a model for transforming bureaucratic irrigation systems.

The author has taken a rather liberal view of what traditional irrigation is in Sri Lanka. The historical roots of various practices appear less important than the fact that they represent something traditional and indigenous, rather than foreign and imposed. Accordingly, they have been a useful tool for reorienting priorities in irrigation development. KOISP most likely will continue to suffer from operation and maintenance problems, but water users' problems and their demands may be given more attention than has been the case in the recent past. The past offers insights — and, indeed, practical advice — for future irrigation management.

References Cited

Abeyratne, Shyamala
    1986   The Development of Institutional Arrangements for Irrigation Water Management in Village Irrigation Systems in Sri Lanka. Paper presented at the Seminar on Irrigation Management and Agricultural Development in Sri Lanka, Agrarian Research and Training Institute, Colombo, Sri Lanka.

Abeywickrema, Nanda
    1988   Government Policy in Participatory Irrigation Management. In *Participatory Management in Sri Lanka's Irrigation Schemes.* Colombo, Sri Lanka: International Irrigation Management Institute.

Alwis, Joe
    1988   Irrigation Legislation and Participatory Management. In *Participatory Management in Sri Lanka's Irrigation Schemes.* Colombo, Sri Lanka: International Irrigation Management Institute.

De Silva, K. M.
   1981   *A History of Sri Lanka*. Delhi: Oxford University Press.
Ekanayake, Ratnasiri, and David Groenfeldt
   1987   *Organizational Aspects of Irrigation Management at Dewahuwa Tank During Yala 1986*. Working Paper No. 3. Colombo, Sri Lanka: International Irrigation Management Institute.
Ellman, A. O., H. A. J. Moll, K. P. Wimaladharma, and J. H. M. Witlox
   1976   *Handbook on Settlement Planning in Sri Lanka*. Peradeniya, Sri Lanka: University of Peradeniya Press.
Farmer, B. H.
   1957   *Pioneer Peasant Colonization in Ceylon: A Study in Asian Agrarian Problems*. London: Oxford University Press.
Gunawardna, R.A.L.H.
   1971   Irrigation and Hydraulic Society in Early Medieval Ceylon. *Past and Present* 53: 3–27.
Harriss, J. C.
   1977   Problems of Water Management in Hambantota District. *Green Revolution*, ed. B. H. Farmer, 364–76. Boulder, Colo.: Westview Press.
Hunt, Robert C.
   1985   Appropriate Social Organization? Water Users Associations in Bureaucratic Canal Irrigation Systems. Revised paper, originally prepared for the annual meeting of the Society for Economic Anthropology, Warrenton, Va., 12 April 1984.
Hunt, Robert C., and Eva Hunt
   1976   Canal Irrigation and Local Social Organization. *Current Anthropology* (17)3: 389–411.
ISPAN (Irrigation Support Project for Asia and the Near East)
   1994   *Participation and Empowerment: An Assessment of Water User Associations in Asia and Egypt*. Arlington, Va.: ISPAN Project Technical Support Center.
Leach, E. R.
   1961   *Pul Eliya: A Village in Ceylon*. Cambridge: Cambridge University Press.
Merrey, Douglas J., and P. G. Somaratne
   1988   *Institutions Under Stress and People in Distress: Institution Building and Drought in a New Settlement Scheme in Sri Lanka*. Occasional Papers. Colombo, Sri Lanka: International Irrigation Management Institute.
Moore, Mick
   1985   *The State and Peasant Politics in Sri Lanka*. Cambridge: Cambridge University Press.
Murphey, R.
   1957   The Ruins of Ancient Ceylon. *Journal of Asian Studies* 16(2):181–200.
Perera, Jayantha
   1988   The Gal Oya Farmer Organization Program: A Learning Process. In *Participatory Management in Sri Lanka's Irrigation Schemes*. Colombo, Sri Lanka: International Irrigation Management Institute.

Perera, K. D. P.
  1986   *Sri Lanka Experience of Integrated Management of Major Irrigation Settlement Schemes (INMAS) Programme on Water Management.* Colombo, Sri Lanka: Department of Irrigation.

Stanbury, Pamela
  1989   *Land Settlement Planning for Improved Irrigation Management: A Case Study of the Kirindi Oya Irrigation and Settlement Project.* Country Paper No. 4. Colombo, Sri Lanka: International Irrigation Management Institute.

Uphoff, Norman
  1986   *Improving International Irrigation Management with Farmer Participation: Getting the Process Right.* Boulder, Colo.: Westview Press.

11

The Relevance of Indigenous Irrigation
A Comparative Analysis of Sustainability

Jonathan B. Mabry and David A. Cleveland

In the latter half of the twentieth century, international agricultural development efforts have invested heavily in irrigation because of its proven abilities to expand cultivated areas and to increase yields on existing croplands. The overwhelming emphasis of these investments has been on what can be described as "industrial irrigation," systems that are developed in industrialized countries and introduced to nonindustrial ones. These systems rely on mechanization, increased energy and capital inputs, large-scale physical infrastructures, centralized management, and fewer crops bred to be more responsive to agrochemicals and increased water supplies.

The many traditional systems of irrigated agriculture that are being replaced can be described as "indigenous irrigation," systems that are locally developed by cultures with unique histories, often over long periods of time, and usually relying on intensive human labor, small-scale water control systems, direct farmer management, and a diversity of crop varieties adapted to local environments. Today, most indigenous irrigators use a mixture of traditional and modern technologies and techniques, and are at least partially linked to national, regional, and global markets. Many indigenous irrigation systems have also been incorporated into larger, state-managed, industrial irrigation systems. Where an adequate degree of local autonomy has been maintained, however, indigenous irrigation systems retain enough of their local adaptations to remain distinctive from industrial systems.

Our purpose in this chapter is to reassess a few conventional wisdoms about the differences between these two types of irrigated agriculture, introduce some new hypotheses, and carry out preliminary tests with some of the available data. Our theory is that indigenous systems of irrigated agriculture are more ecologically and socially sustainable over the long term. Three general hypotheses based on this theory are discussed in the following sections: On average, indigenous systems of irrigated agriculture are (1) more efficient in the use of energy, capital, and natural resources; (2) have more stable yields over the long term; and (3) are more equitable in terms of opportunities, benefits, and risks. We test these hypotheses by reviewing comparative data on relative efficiency, stability, and equity (basic properties of agroecosystems, and essential components of sustainable economies), and suggest some additional data that may be needed to fully assess them. If comparisons among existing data and from future investigations support acceptance of these hypotheses and more specific versions, then we can conclude that indigenous irrigation systems hold useful lessons for the development of sustainable agriculture.

Efficiency

Comparative data indicate that in both nonirrigated and irrigated agriculture the efficiencies of the primary factors of agricultural production are inversely related, there are limits to their substitutability for each other, and the relative returns to the substituting factors of industrial agriculture decline over time. These patterns and trends lead us to hypothesize that indigenous irrigated agriculture is less labor efficient than industrial irrigated agriculture, but makes more efficient use of energy, capital, and natural resources. In the following we show how these differences are obscured by simplistic conventional economic calculations, but are obvious when true values and additional relevant variables are included.

Hidden Costs of Conventional Economic Efficiency

When applied to agriculture, efficiency, as defined by conventional neo-classical economics in terms of benefits relative to costs, discounts long-term returns from soil, water, and other agricultural resources in favor of short-term yields and profits (Norgaard and Howarth 1991). An example is the practice of pumping groundwater at rates in excess of natural recharge rates, as is done on one-fifth of the irrigated land in the United States (Brown and Young 1990). Also, many costs are not included in benefit-cost accounting of agriculture and irrigation. Economic costs are hidden by subsidies, and environmental and social costs are either considered too long-term to be relevant, are removed as "externalities," or are not even acknowledged (Daly and Cobb 1989).

The total costs of industrial irrigated agriculture are often hidden by government use of general tax revenues to subsidize between half and all of the costs of construction and water delivery on large irrigation projects (Sagardoy 1982; Repetto 1986; World Bank 1992). In the 1980s, all of the costs of construction were subsidized on government irrigation projects in Indonesia, Malaysia, Vietnam, Australia, Peru, Saudia Arabia, and South Africa, while between 80 and 100 percent of all water costs were waived in government irrigation schemes in China, India, Pakistan, Indonesia, Mexico, Bangladesh, Egypt, and the Philippines. An average of 60 percent of construction costs, and 60 to 99 percent of water costs, were subsidized on federal irrigation projects in the United States during that decade. This makes industrial irrigated agriculture appear more economically competitive than it would be if the benefiting farmers had to assume complete costs at true market values.

Soil salinization, cropland loss, water contamination, population displacement, increasing incidence of disease, and other long-term costs of industrial irrigation are usually not adequately estimated, but are beginning to outweigh the short-term benefits on a global scale. Salinization resulting from overirrigation and poor drainage has significantly reduced yields on more than 60 million hectares world-wide, about a quarter of the total area irrigated, and 25 million hectares have been abandoned because of salt accumulation (World Resources Institute 1992). Annually, for every additional hectare brought into production by new irrigation schemes, another goes out of production because of salinization (Umali 1993). "Superdams" constructed on major rivers in Africa since 1950 have displaced more than a quarter of a million people, and destroyed their traditional subsistence base of flood recession agriculture (Scudder 1989). It is estimated that Pakistan's Kalabagh Dam will uproot 234,000 people, and adversely affect the water quality and supply of fifteen to twenty million more (Gazdar 1990). Resettlements of about a million people each will be necessary to complete dams on the Yangtze River in China (Ryder 1988) and the Narmada River in India (Alvares and Billorey 1988). Associated with construction of large reservoirs and permanently flowing canals are increased incidence rates of diseases and infections spread by water-borne parasites and other vectors, including malaria, schistosomiasis (bilharzia), liver flukes, filariasis, onchocerciasis (river blindness), dengue fever, yellow fever, chikungunya fever, and encephalitis (Sheridan 1984; Oomen, deWolf, and Jobin 1988). Around the world, a wide variety of successful indigenous modes of subsistence, including irrigated farming systems, are being replaced by industrial irrigation at enormous environmental, social, and health costs that are seldom included in long-term cost projections.

Limited Substitutability of Factors of Production

The neoclassical economic model of agriculture considers land, labor, and capital as the primary resources or "factors" of production, assumes that each can be almost infinitely substituted for any other, and focuses on their changing relative costs in a market-driven search for economic efficiency measured in terms of productivity and profit. The elegant simplicity of this conventional model is also its greatest weakness. It collapses within the single category of capital equally important factors of production such as technology, while cultural knowledge of environments and farming techniques are ignored altogether. It also subsumes energy and other natural resources within the category of land, reduces the contribution of land to rent (and rent to a surplus that is excluded from price determinations), and assumes that if production falls due to reduction in the quality or quantity of land, that production can be maintained or increased by substituting capital or labor (Daly and Cobb 1989).[1]

Daly (1990) proposes that the distinction between natural capital and human-made capital is important, and that their substitutability for each other is ultimately limited. For example, because harvest rates cannot exceed natural regeneration rates to maintain sustained yields, and rates of waste emission cannot overwhelm the absorption capacities of ecosystems, these regenerative and assimilative capacities of ecosystems should be treated as forms of "natural capital," as the failure to maintain them represents unsustainable capital consumption. Capital and labor are relatively substitutable for each other because they both function to transform resources into products, but they cannot significantly substitute for natural capital because the materials transformed and the tools of transformation are complements in their production roles, not substitutes (Daly 1990; see also El Serafy 1991).

Inverse Efficiencies of Factors of Production

In the irrigated form of industrial agriculture, capital and energy are substituted for labor and/or water, with the costs of substituting resources increasing as the costs of those being substituted for decrease. Data from studies of irrigation in the western United States during the late 1970s and early 1980s (Batty and Keller 1980; Roberts, Cuenca, and Hagan 1986) show that decreases in labor or water costs tend to be balanced by an increase in capital and/or energy costs. For example, if labor-intensive hand-moved sprinklers are compared to labor-efficient drip lines, each hour of labor saved per hectare corresponds with a savings of 22.5 cubic meters of water and .346 gigajoules of energy, along with higher operating costs of $9.69 (when installation costs are included, these inverse correlations are magnified until investments are paid off). The data indicate that increased efficiencies of one kind are largely offset by decreased efficiencies in the other factors of production.

In contrast to these trends in industrialized countries (see also Ruttan 1984; Barlett 1989), the diffusion of the industrial system of agriculture to other parts of the world has boosted land productivity only at the costs of increased labor, water, and energy. In regions where capital-intensive mechanization was not affordable for farmers, the necessary higher applications of water, fertilizers, and pesticides associated with adoption of modern crop varieties has led to higher labor inputs to realize the same yields. Studies of twenty different regions in South Asia, Southeast Asia, and Indonesia conducted in the early years of the Green Revolution, between 1966 and 1975, found that paddies planted in modern rice varieties required an average of 22 percent higher labor per hectare than did fields growing traditional varieties (Barker and Herdt, with Rose 1985:127). In these regions the increased land productivity made possible by Green Revolution technology was offset by decreased productivity per unit of labor. This is the same kind of agricultural "involution" described by Geertz (1963) for the intensification of wet rice production in Indonesia prior to the Green Revolution, which raises some important questions about the real-versus-predicted effects of the Green Revolution on labor efficiency.

Declining Returns to Industrial Inputs

Even in regions where efficiencies of one or more production factors have increased, declining economic returns to capital and energy inputs—the substituting factors of industrial agriculture—are becoming apparent. Marginal production response to fertilizer applied to a U.S. cornfield or

Indonesian rice paddy is only half as much as twenty years ago (World Resources Institute 1990). Yields of modern wheat varieties have not changed since 1970 in Pakistan's Punjab region (despite an increase in fertilizer use from 40 to 114 kilograms per hectare), while the marginal increase in grain production from each kilogram of fertilizer applied has fallen to less than 5 to 1 in other parts of Pakistan, compared to over 10 to 1 in the first years of the Green Revolution (Byerlee 1990). After some initial success, pesticides have also lost their effectiveness. Before the use of modern agrochemicals, about a third of the world's annual harvest was lost to insect pests and weeds. Today, while the numbers of pest species known to be resistant to insecticides and herbicides have increased a hundred-fold, the proportion of the crop lost is about the same or greater (Dover 1985). As per capita grain production has leveled off, so have economic returns to the crop varieties, inputs, and methods of industrial agriculture in some of the world's most densely populated countries, including India, Indonesia, Taiwan, South Korea, Japan, China, and Mexico (Brown and Young 1990). Some see this as an indication that the period of dramatic increases in food production due to the Green Revolution may be over, while the costs of industrial agriculture continue to climb and the supporting fossil fuel reserves rapidly dwindle (Brown and Young 1990; Cleveland 1990).

Ratios of Total Energy Outputs to Inputs

Aside from the problems with how benefit-cost ratios are calculated, the difficulty of measuring long-term human and environmental costs, and the inaccuracy of the assumption of unlimited substitutability of production factors, ecologists' calculations of total energy inputs and outputs show that the energy efficiencies of industrial agricultural systems are often negligible, and have declined through this century. Time-series analysis of total energy output relative to the sum of energy inputs (direct fuel and electricity costs plus indirect energy costs of infrastructures, labor, and manufactured inputs such as agrochemicals) in U.S. agriculture between 1910 and 1988 shows that indirect fossil fuel use dominated direct use, with energy efficiency declining rapidly between 1910 and 1973 as total energy use increased more than 500 percent, and then increasing efficiency between 1974 and 1988 as total energy use decreased 42 percent (C. Cleveland 1991).[2]

World-wide comparisons, meanwhile, show that the impressive yields of industrial agriculture, derived from intensive use of manufactured, fossil fuel-based inputs, represent lower energy efficiency relative to less industrialized systems of agriculture (Cox and Atkins 1979; Pimentel and Pimentel 1979). This contrasting pattern is also apparent in comparisons of ratios of energy outputs to inputs in different irrigated systems (table 11.1). An irrigator in the Ziz Valley oasis of southern Morocco, for example, invests sixty-four times more manual labor than his California counterpart to produce a wheat yield of up to 4,000 kilograms per hectare, but he uses less than half as much petroleum inputs in the forms of agrochemicals and fuel for machinery (Mabry, Ilahiane, and Welch 1991). The Moroccan's crop yield thus represents a ratio of energy return of up to 12.3 to 1, while the Californian's average crop yield of 4,640 kilograms per hectare represents a relatively meager energy return of 7.3 to 1 (Pimentel and Pimentel 1979). Maize crops produced from raised and drained *chinampa* fields in the Valley of Mexico represent an impressive return of 103.4 to 1 (Sanders and Santley 1983), while irrigated rice production in California has one of the lowest energy output-to-input ratios in world agriculture: 1.4 to 1 (Pimentel and Pimentel 1979). Traditional systems of flood recession agriculture, which rely on "natural irrigation," also have relatively high potential energy efficiencies: up to 102.5 to 1 for riverbank maize production in southern Belize (Wilk 1985); 71.7 to 1 for floodplain rice production along rivers in the upper Amazon Basin in Peru (Chibnik 1990); and 152 to 1 for sorghum cropping on the alluvium of the Senegal River (Park 1992).[3]

Economies of Scale in Irrigation and Agriculture

Increasing farm sizes and a search for economies of scale are also significant trends in the development of industrial agriculture. Here again, direct and indirect state subsidies of production costs and discounting of future returns distort both market processes and measurements of agricultural efficiency. For example, in the western United States, despite the provision in the 1902 Reclamation Act limiting use of water from federal water projects to 160 acres (65 hectares) per farmer, or 320 acres (130 hectares) per married couple, farmers were at first able to avoid breakup of their large holdings by irrigating the legal amount with federal water and pumping free groundwater for the remainder (Worster 1985). By mid-century, though, the provision was interpreted more loosely by the Bureau

Table 11.1 Energy inputs and outputs in indigenous and industrial irrigated grain

Grain and Type of agriculture	Location	Human labor		Animal labor	Fertilizers[1]		Fossil-fuel based[2]	Total energy input
		hrs/ha	kcal/ha[a] (×000)	kcal/ha[b] (×000)	kg/ha	kcal/ha[c] (×000)	kcal/ha (×000)	kcal/ha (×000)
Maize								
Recession	Southern Belize	448	67.2	—	—	—	—	67.2
Floodwater	Central Mexico	100	25.0	594.0	—	—	—	619.0
Canal irrigation	Central Mexico	153	78.8	594.0	—	—	—	672.0
Sprinkler irrigation	Western U.S.	12	5.6	—	280	2,225.6	6,526.5	6,532.1
Raised-drained field	Central Mexico	200	103.0	—	—	—	—	103.0
Wheat								
Recession	Egypt, pre-Aswan Dam	555	83.2	1,926.0	—	—	—	2,009.2
Canal irrigation	Ziz Valley, Morocco	449	231.2	—	95	807.5	848.9	1,080.1
	Eastern Jordan Valley	520	267.8	—	34	289.0	330.4	598.2
	Uttar Pradesh, India	615	316.7	1,926.0	—	—	41.4	2,284.1
Sprinkler irrigation	Western U. S.	7	3.2	—	106	841.0	2,112.4	2,115.7
Rice								
Recession	Upper Amazon, Peru	634	95.1	—	—	—	—	95.1
Canal irrigation	Lowland Philippines	1,864	960.0	952.0	30	280.0	321.4	2,276.9
	Eastern China	1,048	539.7	952.0	195	3,616.3	1,639.9	3,131.7
	Lowland Japan	1,128	524.5	—	431	1,639.9	6,721.7	7,246.2
	Western U.S.	12	7.9	—	347	2,859.4	14,431.6	14,439.5
Sorghum								
Recession	Senegal Valley, Mauritania	300	45.0	—	—	—	—	45.0
Sprinkler irrigation	Western U.S.	12	5.6	—	119	1,000.8	5,372.6	5,376.1

(1.) Nitrogen, phosphorous, potassium
(2.) Machinery, fuel, fertilizers, pesticides, drying, transportation
(a.) Recession: 150 kcal/hr; floodwater: 250 kcal/hr; canal irrigation/raised-drained field: 515 kcal/hr; industrial canal/sprinkler irrigation: 465 kcal/hr
(b.) Animal labor: 3000 kcal/hr
(c.) Nitrogen: 14,700 kcal/kg; phosphorus: 3000 kcal/kg; potassium: 1600 kcal/kg.

production, 1975–1985.

		Outputs		
Grain yield	Labor productivity	Total energy output	Energy return ratio
kg/ha	kg/hr	kcal/ha*d* (× 000)	kcal output/input
230–1,940	.5–4.3	816.5–6,887.0	12.1–102.5
800–1,000	8.0–10.0	2,840.0–3,550.0	4.6–5.7
1,000–1,400	6.5–9.1	3,550.0–4,970.0	5.3–7.4
5,390	449.5	19,134.5	2.9
3,000	15.0	10,650.0	103.4
570–2,850	1.0–5.1	1,892.0–9,462.0	.9–4.7
1,346–4,000	3.0–8.9	4,468.7–13,280	4.1–12.3
2,500–4,500	4.8–8.6	8,300.0–14,940.0	13.9–25.0
1,600	2.6	5,312.0	2.3
4,640	562.9	15,404.8	7.3
2,000	3.1	6,820.0	71.7
2,050	1.1	6,990.0	3.1
3,770	3.6	12,885.7	4.1
5,710	5.1	19,471.1	2.7
6,160	362.3	21,005.6	1.4
400–2,000	.5–2.5	1,368.0–6,840.0	30.4–152.0
3,030	252.5	10,362.6	1.9

(*d.*) Maize: 3550 kcal/kg; wheat: 3320 kcal/kg; rice 3410 kcal/kg; sorghum 3420 kcal/kg.

Sources: Pimentel and Pimentel 1979; Wilk 1985; Sanders and Santley 1983; Stanhill 1984; Barker, Herdt, and Rose 1985; Chibnik 1990; Qasem 1990; Mabry, Ilahiane, and Welch 1991; World Resources Institute 1992; FAO 1991; Park 1992

of Reclamation, when the 160-acre allotment was extended to every adult member of a farm household, and time limits on leasing were ended. In this case, an economy of scale was obtained by the larger farms of the western United States through greater use of subsidized water from federal projects, unregulated individual use of common property water resources, or both.

In addition to the distorting effects of subsidies and discounting, the pattern found through cross-cultural comparisons challenges the economies-of-scale model by suggesting that productivity (output per unit of land area over time) and efficiency (ratio of output per unit of input over time) may be inversely related to the scale of production units (sizes of fields and farms). This possibility is usually overlooked in industrial countries due to the belief that smallholder production is less efficient than large-scale agribusiness, and in developing countries due to the need to rapidly boost national productivity. According to studies in India, however, irrigated agriculture increases in productivity as the size of farms decreases (Saini 1979). World-wide, in fact, there is an inverse relationship between area cultivated and yield in all types of agriculture; on average, smallholders get more out of the same amount of land (Berry and Cline 1979; Netting 1993). Strange (1988) has concluded from comparative data that, even in the United States, where farmers take pride in the labor efficiency of their large operations, smaller and medium-size farms are more efficient in resource use, their production costs relative to gross sales are lower, and their profits as a percent of gross sales are higher.[4]

Management Structure
and Administrative Efficiency

Comparative data suggest that, in addition to the scale of production units, efficiency and productivity in irrigated agriculture is related to management structure, or the locus of resource control. Because their administration is less efficient in terms of information flow and decision making, centrally managed systems have lower productivities than locally managed ones. Ostrom (1993), in her comparisons of 108 irrigation systems in Nepal, found that farmer-managed systems averaged crop yields of 6 metric tons per hectare, compared to 5 metric tons per hectare on agency-managed systems. The tendency of centrally managed systems toward lower productivity is probably related, at least partly, to the logistical difficulty of efficiently monitoring and directing multiple local units within large-scale systems from single, distant locations.

Efficiency and
Irrigated Agricultural Production

The studies cited and our comparisons seem to support our hypothesis concerning efficiency in irrigated agriculture. The data show inverse trends in the efficiencies of different agricultural resources in the transformation to the industrial mode of agricultural production, declining returns over time to the substituting factors of industrial agriculture (capital and energy), and relatively low efficiency in terms of the ratio of total energy outputs to inputs in industrial irrigated agriculture. The data also indicate that smaller farms and locally controlled irrigation systems tend to be more efficient and productive than larger farms and centrally controlled systems.

We suggest that these patterns indicate limitations to the substitutability between primary factors of agricultural production, highlight how agricultural efficiency is measured in conventional economic models, and challenge the applicability of economy-of-scale models to farming. This does not mean, however, that industrial irrigated agriculture is less productive over the short term, or is inefficient in terms of labor or even water use in the case of some application systems; we are only pointing out its hidden costs, tradeoffs, and negative long-term rates of return that are not usually considered. More complete accounting, along with a comparative perspective, will allow further testing of our hypothesis that the kinds of efficiency found in indigenous systems of irrigated agriculture, including those kinds not usually measured, are related to their appropriate technologies, small scales, and local institutions for resource control.

Stability

In agriculture there is evidence for a causal and inverse relationship between stability and productivity at the biological, ecological and sociocultural levels, which is mediated, to a large degree, by diversity (Cleveland 1993, 1995). Because indigenous agriculture tends to be more biologically, ecologically, and socially diverse than industrial agriculture, it is logical that the replacement of indigenous irrigation by industrial irrigation, as part of the industrialization of agriculture, increases instability. This deduction, supported by comparative data, leads us to hypothesize that industrial irrigated agriculture has less stable yields than indigenous irrigated agriculture. Initially, it may seem illogical to suggest that supplying water by irrigation to crops formerly dependent only on unreliable rainfall would decrease yield stability. As we hypothesize, however, it is not irrigation per se that increases instability, but rather the tendency of industrial irrigation to increase irrigation intensity, diminish biological diversity, and decrease local management that account for the tendency of irrigation to decrease yield stability.[5]

Irrigation Intensity

The intensity of irrigation, measured as the amount of water applied in a given area over a given time period, which includes expanding irrigated lands in a country or region, tends to increase with industrialization. This usually involves an increase in cropping intensity and a shift to more water-consumptive crops, both of which raise the demand for water per unit area, as well as demand for other inputs such as chemical fertilizers and pesticides. Comparative data suggest that, rather than the irrigation per se, the increased use of these related inputs and of water-responsive modern crop varieties decreases yield stability.

Most quantitative analysis of yield stability has been done on short-term variability and on comparison of periods before and after the development or introduction of industrial agriculture (Anderson and Hazell 1989). In one of the first reports to specifically examine yield stability in relation to irrigation, it was argued that "there is no evidence to support" the assertions made in the Green Revolution literature that modern technology, including industrial irrigation infrastructures, leads to increased yield stability (Barker, Gabler, and Winkelmann 1981:74). Comparative data were used to show that, although irrigation may potentially reduce moisture stress, it is frequently associated with an intensification of crop production and input use that is "destabilizing" (Barker, Gabler, and Winkelmann 1981:63). In a shift to industrial wheat production producing 63 percent higher yields, for example, irrigated farms in the Yaqui River valley of Mexico experienced a 45 percent increase in yield variability (measured by standard deviation), to a level higher than that measured for rainfed industrial wheat production in Nebraska, United States

(Barker, Gabler, and Winkelmann 1981:59). A comparison of data from East Asia with the evidence from South and Southeast Asia supports this general hypothesis of an inverse relationship between industrialization of irrigation and yield stability. Barker, Gabler and Winkelmann (1981:64–65) conclude that the data show higher absolute variability but about the same range of relative variability in East Asia, which had "the most highly developed infrastructure and the highest yields" of rice.

Many other researchers have identified the same trend.[6] Mehra (1981) has compiled data on variability in food grain yields in rainy (*kharif*) and dry (*rabi*) seasons in India, revealing that variability tends to be higher during the rainy season, when a smaller proportion of the cultivable area is irrigated. She interprets this as showing that "irrigation by itself appears to reduce yield variability," but "when irrigation is combined with intense input use, yield increases, but so does variance" (Mehra 1981:30). She also found that the absolute variation (standard deviation) for all crops increased by 75 percent, and for food grains by 65 percent, from the period before the Green Revolution, when local crop varieties predominated.

Although confirming Mehra's findings, Hazell's (1982) reanalysis suggests that the nearly 50 percent increase in relative instability between pre- and post-Green Revolution periods, from 4.03 to 5.85, was due primarily to increasing covariation of yields of different crops in the same state and between states, rather than increases in crop yield variances per se. In either case, it is the synergistic movement toward modern crop varieties more responsive to increased supply of water and other inputs and the intensification of production and inputs (both typical of industrial irrigation) that leads to increased instability.[7]

Water Supply

Increasing irrigation intensity is associated with increasing exploitation of water supply, as more sources are tapped at greater rates of use. Thus, the stability of water supply is a major factor affecting yield stability. Mehra (1981:37) believes that the evidence from her study of variability in Indian food grain production shows that "when seed-fertilizer technology is combined with assured irrigation, the tendency for variance to increase is neutralized." In the real world, however, farmers seldom receive the optimal amount of water at the right time to obtain the potential yields of industrial crop varieties. For example, Walker (1989) has shown that in

India, irrigation tends to increase covariance between regions for sorghum yields, but leads to reduced interregional covariance for millet yields. He concludes that "this puzzling result could stem from the fact that irrigated pearl millet often entails only one or two applications of water and is largely cultivated where water supply is most uncertain" (Walker 1989: 99). Variability in water supply, then, may explain this pattern for millet.[8]

As more accessible surface water supplies are allocated, irrigation systems tend to exploit groundwater, often at unsustainable rates, thereby stabilizing yields over the short term, but only by decreasing long-term stability. With fossil fuels, groundwater is pumped from deep tubewells — an industrial technology widely adopted in nonindustrial countries — to provide almost all of Libya's and Saudi Arabia's water supplies, 95 percent of Tunisia's, and 75 percent of Israel's and Iran's (Postel 1989). By the mid-1980s, tubewells supplied 40 percent of the irrigated area in Bangladesh, and falling water tables and resulting saltwater intrusion has raised pumping costs and reduced water quality (Mandal 1987). Because of overpumping, groundwater levels are falling between 1 and 2 meters per year in parts of northern China, and 2.5 to 3 meters per year in the southern Indian state of Tamil Nadu; more than 4 million hectares in the United States (about 20 percent of the total irrigated area) is supplied by pumping in excess of natural recharge rates (Postel 1992).

Biological Diversity

An important corollary of increasing irrigation intensities and increasing rates of water resource exploitation is often the substitution of more water-consumptive crops and modern crop varieties for local farmers' or folk crop varieties.[9] In fact, the spread of the Green Revolution in the Third World has largely been limited to the irrigated zones, and modern crop varieties have not performed well in marginal, rainfed areas of Asia, Africa, and Latin America, where more dependable local varieties still predominate (Chambers 1984a; Barker and Herdt, with Rose 1985). Decreasing crop genetic diversity may also contribute to destabilization of irrigated agriculture. For example, among ten Asian countries analyzed for before- and after-Green Revolution periods, "there was a tendency for the percent of area in modern varieties, yield, standard error, and coefficient of variation to increase with a rise in the percentage of irrigated rice area" (Barker, Gabler, and Winkelmann 1981:63). Plant breeders and agronomists often find that crop yield cannot be increased without decreasing the stability of yield when the crop is exposed to drought, waterlogging, pests and pathogens, and other stresses. Although few comparisons between folk varieties and modern varieties have been carried out, existing data suggest that modern varieties often have higher yields than folk varieties under optimal conditions, while heterozygous and heterogeneous folk varieties with broad resistance to a variety of stresses often have higher production than modern varieties under stress (Cleveland, Soleri, and Smith 1994).

It is not only through decreasing biological diversity that the spread of modern varieties may destabilize irrigated agriculture, but also through the high costs of other necessary inputs. Pandey (1989:236) notes that, because irrigation is often part of a package that includes modern varieties and fertilizers, it "increases the marginal productivity of other complementary inputs," leading to "more intensive cropping practices." In the case of modern varieties, this means that Third World farmers must often buy a whole package of industrial inputs from distant sources. Uncontrollable interruptions of supply, along with difficulties in obtaining credit and unpredictable variations in annual incomes, also lead to yield instability when agricultural inputs must be purchased rather than locally obtained.

At the ecological level, beginning in the field, the diversity of traditional agriculture in the form of many crops and many varieties of each crop, and diversity in soil conditions and field locations may often result in higher yield stability than demonstrated in uniform, industrial agroecosystems. Intercropping tends to increase yield stability (Lynam et al. 1986), and, although there are some exceptions (for example in China), irrigation leads to a significant increase in monocropping (Jodha 1990). The spread of uniformity across the landscape in the form of similar crops and crop varieties, planting patterns, inputs, and government agricultural price supports, all of which usually accompany the spread of intensive industrial irrigation, tend to increase the covariation or synchrony between crops and between regions. Along with greater instability of modern varieties per se under stressful conditions, this covariation contributes to yield instability (Anderson and Hazell 1989, Hazell 1989).

Management Structure

At the social level, there is support for the hypothesis that centralized management structure causes an increase in yield instability by removing the ability and power of farmers and local communities to manage irrigation in response to changing local conditions. A number of social scientists investigating irrigation systems believe that increasing centralization of water distribution leads to a process of positive feedback that drives a centralizing trend to the point where production is pursued in almost complete isolation from environmental and local level social pressures, leading inevitably to a system crash (Merrey 1987; Swearingen 1987; Chambers 1988:239–242; Cleveland 1996, Uphoff, Wickramasinghe, and Wijayaratna 1990; Wade 1986). In international agricultural development, the failure of projects planned and administered in ignorance of local systems, as most have been, is legend.

In systems that centralize water control in the hands of headenders, the problems of tailenders resulting from decreased adequacy, reliability, and timeliness of water supplies are well-known, but Chambers describes a frequent, but less-recognized, problem among headenders in Asian irrigation systems: "Quite often, headreach farmers appear to be locked into their own variety of the tragedy of the commons. This is especially marked with field to field irrigation of paddy. The abundant issue of water and consequent flooding, combined with the cultivation of paddy by his neighbors, remove any option from a farmer to grow anything but paddy; and then because all farmers follow this practice, waterlogging, salinity, and flooding ensue, reducing or eliminating yields" (Chambers 1984b:41).

Community management of common property resources such as irrigation water sources and delivery systems may increase yield stability by managing for long-term conservation (sustainability), and these common property management institutions are more likely to evolve and persist where viable, local communities with control over local resources already exist. A review of a number of independently conducted case studies of common pool resource management supports the theory that irrigation water and other common property resources tend to be managed by local community groups such as water users' associations when there is a common understanding of problems and alternative solutions, when decision-making costs are less than benefits, and when local organizations are nested in a hierarchy of organizations that protects them from external forces such as government interference, market fluctuations, and population pressure (Ostrom 1992, 1993).

The effect on stability of industrial irrigation in increasing central management and simultaneously reducing local control is illustrated by the Salinity Control and Reclamation Projects (SCARPS) of South Asia. SCARPS are attempts to salvage large-scale industrial irrigation systems from production problems by further application of the same central management industrial development approach that created the problems in the first place. SCARPS are dominated by engineers who see "physical, hardware remedies . . . as the *only* remedies," and completely ignore obvious solutions, such as supplying less water (Chambers 1988:78, emphasis in original). For example, in the Swabi SCARP in northwest Pakistan, water-user associations and a demand system were proposed as ways to increase production by increasing farmer participation, but their design and application tend to decrease participation and increase instability (Cleveland 1996). In fact, the main purpose of water-users' associations seems to be to force the irrigators to carry out the program planned for them by the central bureaucrats. As with most conventional irrigation development in Pakistan, this project also ignores farmers' agricultural expertise and their knowledge of the irrigation system.

Stability and Irrigated Agricultural Production

Overall, the available data do appear to support our hypothesis that the increasing intensity, decreasing biological diversity, and centralization of management that accompany the industrialization of irrigation tend to increase variability in yield compared with indigenous irrigation. Although the effect of irrigation on yield stability depends on the timing, season, and predictability of the water supply, irrigation generally increases absolute variability because of the associated increases in yield, planting of water-responsive modern crop varieties, homogeneity of the growing environment, and covariance between neighboring fields. Efforts to assure the reliability of irrigation water over the short run, such as increased extraction from the source (larger dams or deeper tubewells, for example), are also likely to lead to larger fluctuations or even failures in the water supply over the long term, due to depletion of aquifers, siltation of reservoirs and canals, waterlogging and salinization of fields, and a centralized administration increasingly inattentive to feedback from local ecological and social conditions.

Despite general support for our hypothesis, the data considered here are not entirely consistent, and no predictive statements are yet possible that apply to specific situations. Yield stability covaries with other factors besides the degree of agricultural industrialization, as for example, in a comparison of data from Bihar and Tamil Nadu states in India (Barker, Gabler, and Winkelmann 1981:63). Our hypothesis predicts that, because yields are twice as high in Tamil Nadu, variability should also be higher — yet the data show the opposite, probably because of the high frequency of severe floods and droughts in Bihar. To accurately compare stability and yield, therefore, environmental disturbances must be experimentally or statistically controlled. Pandey (1989:241) also points out that the evidence for the effects of irrigation on production variability is inconclusive because the conventional production function treats irrigation as a constant and homogeneous input, when in fact the effect of irrigation depends on the "quantity applied, the timing of application, stability of water supply, water distribution rules, plant characteristics" and relationships that are "dynamic, interactive, and stochastic." There does appear to be sufficient evidence, however, to cast doubt upon the conventional wisdom that yields are stabilized through industrialization of irrigation, including increased irrigation intensity, adoption of modern crop varieties, and central control over water and other inputs.

Equity

In irrigation systems, the equity of water allocation can be assessed in terms of how close the actual distribution fits a culturally defined ideal such as temporal priority of use, proportionality to land holdings or labor contributions, or variation among users in areas irrigated, field soil moistures, crop yields, or farming incomes. In farmer-managed irrigation systems, equity is often measured in terms of risk-spreading or wealth-leveling among system users. But cultural differences mean that equity is inherently subjective; in many places, there are rules for regulating use of water resources that reflect cultural concepts of equity, yet at the same time result in advantageous arrangements for some socioeconomic groups or classes. Indeed, established inequalities are often reinforced by the customary rules of resource access, or by local political structures. Differential access to naturally flooded land in Somalia maintains the stratification of power (Besteman, chapter 3). In some areas of Mexico, water is allocated according to the political connections of local bosses, who thereby increase their power (Yates 1981).

Clearly, equity—like beauty—lies in the eye of the beholder; it cannot be measured objectively by a single personal or cultural criterion. Yet, comparative data leads us to hypothesize that, whether equity within a group of irrigators is measured in terms of relative evenness in access to the means of production (irrigable land, water supplies, delivery systems), obligations for contributions to construction and maintenance (labor, capital), distribution of benefits (yields, profits, employment), or shares of risk (water shortages, pest infestations, and so on), indigenous irrigated agriculture tends to be more equitable than industrial irrigated agriculture.

Contributions and Benefits

In addition to rules specifying use rights, rules regarding the relationship between labor or capital contributions and the distribution of benefits also determine the relative equity of an irrigation system. For example, in a system that assigns equal shares of water to each household but requires all males to contribute labor, larger households have to contribute a disproportionate amount of labor; this outcome can be avoided by allocating water in proportion to labor inputs (Ostrom 1993). Irrigation systems with unequal distributions of benefits do not function as well as systems in which benefits are distributed in ways considered equitable by a majority of users. For example, systems that result in unequal distribution of benefits have lower rates of compliance to rules. In a comparison of forty-three case studies of community irrigation systems, Tang (1992) found that systems with higher variance in average annual family incomes demonstrated lower degrees of rule conformance and fewer contributions to maintenance.

Headender-Tailender Relationships

The relative positions of irrigators within the hierarchical management structures of all but the smallest irrigation systems influence the equity of their shares in system benefits. In some areas of southern India, farmers at the head end of the system not only apply more water to their fields than necessary, but also interfere with the main canal system to maximize their supplies, thereby reducing the flow to farmers farther down the canal (Wade 1986). Where headenders need the capital and labor of tailenders, however, relatively more water reaches the tail of the system, while modernization of headworks often has the effect of decreasing the equity of water distribution because the contributions of tailenders are no longer needed for maintenance.

Ostrom (1993) has found that, in negotiations over the rules of water use between the members of a local water-user association, the bargaining power of tailenders is greater if their labor is needed to maintain the system. On the other hand, because the contributions of irrigators from all parts of the system are no longer needed, many successful farmer-organized water-user associations collapse soon after their systems have been "modernized" to decrease labor requirements and maintenance costs. In a comparison of 127 irrigation systems in Nepal, it was found that uneven distribution of water between head and tail ends was correlated with installation of permanent headworks in agency-managed systems, which no longer required the labor of tailenders to maintain (Ostrom, Benjamin, and Shivakoti 1992). During the wet season, adequate water reached the tail ends of only half of the agency-managed systems, compared to 90 percent of the farmer-managed systems. During the dry season this pattern was even stronger, with adequate water reaching the tail in only 8 percent of the modernized systems, compared to 25 percent of the traditional systems.[10]

In some cases, tail-to-head distribution of irrigation water leads to increased equity. Netherly (1987) has described the relative equity of the pre-Hispanic practice of tail-to-head distribution in Peru, compared to the less equitable system of upstream-user priority imposed during the Spanish colonial period. From his comparisons, Chambers (1984b) concludes that redistribution of some water from head to tail can potentially achieve several important objectives simultaneously, including increased equity.

> If less water is issued at the top, farmers there can grow crops that are more suitable for the soil, and if water is redistributed to the tails, then total production should rise, and equity will be served. Stability will be enhanced through reduced waterlogging. Carrying capacity will be

increased through higher labor demand both in the head reaches . . . and in the tail where irrigated area and intensities increase. Well-being should gain through these effects, through reduced health hazards from standing water in the head reaches, and through more canal water for domestic purposes in the tails. (Chambers 1984b:37)

Variability in Landholdings and Incomes

The gap between the largest and smallest landholders is frequently magnified by modernization, or industrialization, of irrigation, and this influences the distributions of water and incomes. In his comparison of the effects of introducing public-managed irrigation in two major irrigation projects in Maharashtra, India, Dhawan (1984) found that all farm incomes increased, but that income differences between large and small farmers also increased due to the former's better access to water, credit, and extension services. In public irrigation schemes in central Tunisia where water is often scarce and therefore costly, farmers who cannot afford irrigation either abandon it or greatly restrict its use, making more water available to the farmers who can afford it, whose yields and incomes thereby increase (Salem-Murdock 1990). A similar pattern has been documented in the Cape Verde Islands (Langworthy and Finan, chapter 8). Chambers (1984b:29) refers to comparative data from several parts of South and East Asia to suggest that "water distribution between farms tends to be more equitable the more equal the landholdings are, quite apart from the direct effect of the relative equality of the farm sizes."

Loci of Resource Control

The locus of control of an irrigation system also affects the equitability of benefit distribution, as well as efficiency and stability. Whether crucial management decisions are in the jurisdiction of local farmers or distant bureaucrats seems to make a difference. Comparisons indicate that income variance among irrigators tends to be less in locally managed systems than in agency-managed ones, suggesting that there is an inverse correlation between centralized control and equity. In twenty-six different cases, Tang (1992) found that income variance was generally higher in agency-managed irrigation systems than in locally managed ones, although half of the community systems displayed moderate income variance among irrigators. Some of this variability, the data indicate, is related to the asymmetrical benefits accruing to headenders from construction of permanent headworks that increase total water delivery only to the top of canal distribution systems.

Equity and Irrigated Agricultural Production

The patterns in the comparative data reviewed here support our hypothesis about equity in irrigated agriculture. The equitability of water allocation can be measured alternatively according to different, culturally defined principles of resource access and use. Among the factors that affect equitability in most irrigation systems are the rules specifying the relationship between contributions and benefits, the relative structural positions of irrigators, variance in landholdings and incomes, and the loci of resource control. Indigenous irrigation systems should not, however, be overly romanticized or intellectualized as ideally egalitarian. Traditional irrigation communities and water-user associations are not representative of Wittfogel's "hydraulic despotism"—they are usually small, simple, and decentralized. They are also not examples of Marx's "primitive communes"—risks are spread evenly through the group, not surpluses or profits. In terms of decision making, irrigation communities and water-user associations may be natural loci of "agrarian democracy" (Netting 1989), but they are not necessarily the egalitarian "village republics" envisioned by some (Wade 1988). Comparisons show that they tend to be internally egalitarian but exclusionary, and previous local inequalities may be reinforced by the marginalization of the less powerful during the process of development.

The Relative Substitutability of Indigenous and Industrial Irrigation

We conclude that the data we have reviewed support acceptance of a number of hypothetical comparisons between indigenous and industrial irrigation based on the theory that indigenous modes of irrigated agriculture tend to be more sustainable (table 11.2). In general, indigenous systems use energy and natural resources more efficiently, have lower but more stable yields, and are more equitable in the distribution of opportunities, benefits, and risks. A capital- and energy-intensive system of irri-

Table 11.2 Hypothetical comparisons between indigenous and industrial agriculture, based on the theory that indigenous systems are more sustainable.

System level	Indigenous	Industrial
Crops		
Number and types of varieties	More folk varieties (FVs)[a]	More modern varieties (MVs)[b]
Genetic diversity	Higher	Lower
Yield under marginal conditions	Higher	Lower
Yield under optimal conditions	Lower	Higher
Yield stability	Higher	Lower
Farms, fields		
Number of crop species, varieties	Higher	Lower
Diversity of environments	Higher	Lower
Size	Smaller	Larger
Outside inputs	Lower	Higher
Specialization	Lower	Higher
Farmer risk	Lower	Higher
Region		
Number of fields, crop systems	Larger	Smaller
Outside inputs	Lower	Higher
Synchrony of yields	Lower	Higher
Irrigation system		
Cropping intensity	Lower	Higher
Water sources	Local, many	Central, few
Control of water distribution	Local	Central
Water consumption of crops	Lower	Higher
Risk of waterlogging, salinization	Lower	Higher
Whole system		
Proportion of NPP[c] used	Lower	Higher
Social organization	Local community	Hierarchical bureaucracy

Table 11.2   (*continued*)

System level	Indigenous	Industrial
Equity	Higher	Lower
Energy efficiency	Higher	Lower
Diversity	Higher	Lower
Stability	Higher	Lower

Source: Adapted from Cleveland 1995
a. Also called "farmers' varieties," "traditional varieties," "landraces"
b. Also called "high-yielding varieties" (HYVs)
c. "Net primary product" of plant photosynthesis

gated agriculture may be able to sustain considerable levels of population by raising yields, but the productive resource base may be degrading. The apparently high levels of efficiency, stability, and equity of industrial irrigation, in reality based on external inputs and central management, may over the long term, be low relative to indigenous irrigation systems that must rely solely on local resources of labor, land, and leadership.

To accurately model the sustainability of an agroecosystem, calculations of efficiency must include the costs of capital, energy, and water subsidies, and also the contributions of forms of capital other than human-made. In the accounting, subsidizing and future discounting of natural capital should be replaced by its adequate valuation, and calculation of its depreciation (Norgaard and Howarth 1991; El Serafy 1991). Ecosystem complexity and genetic diversity should be included in this category, along with ecosystem regenerative and assimilative capacities. The concept of capital, even if restricted to the neoclassical sense of the economy's stock of real goods that are capable of producing further goods and utilities, should also be extended to include human organizational capacities such as social capital (Coleman 1988). Examples include those institutions in local irrigation organizations that facilitate sustainable resource management, equitable allocation, and stable production by providing collective-action mechanisms for rule-making, compliance-monitoring, conflict resolution, decision enforcement, and rule formulation and modification (Ostrom 1993; Mabry, chapter 1).

Even with these adjustments, however, the neoclassical economic model may be of limited applicability to agriculture, including irrigation. It has been said that "rivers express a rationality different from economics"

(Worster 1984:58). The same can be said of traditional, subsistence-oriented systems of irrigated agriculture, which often function in contexts not dominated by market economic forces, and which measure efficiency in terms other than productivity and profit.

Although the data reviewed here offer some support for our hypothesis, our comparisons between indigenous and industrial systems of irrigation agriculture summarize only a few of the complex relationships between the many variables involved in agricultural production, and the results may often depend on the specific sets of cases considered. There is certainly much more to learn, and new data must be collected to test these hypotheses based on the theory of the greater sustainability of indigenous types of agriculture. At the very least, a transformation in how we think about and measure the properties that determine agricultural sustainability is necessary to help redirect irrigation development — a change in course demanded by ecological, economic, and demographic realities.

Because lending for water development projects by international donors has declined by more than 60 percent in the last decade (Postel 1989) as world population has continued to explode, governments in the developing world will be forced to shift their efforts to rehabilitating indigenous irrigation systems to grow food crops, rather than replacing them with expensive, complicated foreign water control technologies to boost cash crop production for distant markets (Coward and Levine 1989). The result of this shift in technology and management may be a "water revolution" that will stabilize food production and deliver benefits to the least-advantaged farmers (Chambers 1980; Freeman 1989). It will most likely be driven by a synthesis of indigenous knowledge and industrial technology, but based on values and goals more similar to those of indigenous than industrial systems of agriculture.

The critical obstacles are, inescapably, the high rate of resource consumption by a relatively small proportion of the global population and the high rate of world population growth. Development of sustainable agriculture may only be possible with a smaller number of people who consume, on average, fewer resources. The final choice among systems of intensive food production may be between a monolithic, industrial irrigated agriculture that can support a large and growing population for only a short time, or a diversity of locally adapted systems of irrigated agriculture, based on indigenous knowledge, that can sustain smaller, stable populations over the long run. In the words of farmer-poet Wendell Berry, after seeing the vast differences between traditional and industrial types of irrigated agriculture in the southwestern United States, "It is better to sustain a small population indefinitely than to build up a large artificial population on an agricultural system of which the basic principle is a willingness to destroy itself" (Berry 1981:67).

Acknowledgments

This chapter has evolved substantially through multiple drafts and has benefitted from comments by Robert Netting, Daniela Soleri, Charlie Stevens, John Welch, and three anonymous reviewers. The authors alone are responsible for its content.

Notes

1. If energy is considered an equally significant factor of production, one of the clearest trends in the industrialization of agriculture is the substitution, for labor and land, of human-made capital (in the forms of machinery, scientifically bred seeds, and infrastructures like barns, silos, greenhouses, and irrigation systems) and energy (in the forms of electricity and fossil fuels for tractors, pumps, and other machinery, and also agrochemicals derived from fossil fuels). In the nineteenth century, capital inputs were emphasized in North America because of the relatively abundant land, large farms, and limited labor supply, while energy inputs were more important in Europe where land was scarcer and labor more plentiful (Dahlberg 1990). Energy use in agriculture has increased in both regions since World War II, with the fastest rates of substitution of fossil fuel for human labor in the United States and the United Kingdom (Stanhill 1984). This same trend has occurred in irrigation at a global scale, as increased use of pumps powered by fossil fuels has accompanied the spread of industrial irrigation. Between 1950 and 1985, as total energy use in agriculture increased about 700 percent, fuel use in irrigation increased 1200 percent (Albertson and Bouwer 1992).

2. Scaling of energy inputs and outputs to the number of acres harvested reveals that output per acre steadily declined relative to energy use per acre between 1910 and 1974 (C. Cleveland 1991). This long downward trend in energy efficiency was caused by accelerating use of petroleum and petroleum-derived agrochemicals. The recent upward trend is due to sharply declining use of agrochemicals after the energy price shocks of 1973–1974 and 1980–1981, and also the removal of marginal, lower-quality land from production under government-subsidized land conservation and price stabilization programs, thereby increasing the average yield per acre.

3. In reality, these ratios are probably even higher where draft animals are grazed on post-harvest stubble and natural pasture; animal labor inputs are based on energy outputs that are otherwise partially lost, or that are external to the cropping system.

4. Studies of irrigated agriculture in industrialized and developing regions have also shown that greater numbers of small holdings on the same amount of land provide more

livelihoods per unit of water, thus improving rural quality of life (Goldschmidt 1978; Chambers 1984a).

5. A central component of agricultural stability is yield stability, or "the relatively constant annual yield of a crop grown by a farmer," and is "one of the most important issues facing world agriculture and food production; in some cases, stability is equally as important as yield itself" (Federer and Scully 1993:612). We define yield stability as a measure of variability of yield through time or across space, measured by variance (absolute stability) or the coefficient of variation (relative stability) (see Cleveland 1993).

6. Although the overall trend of increasing instability does appear to be accepted by most researchers, there are certainly many exceptions, and conclusions depend on research designs (e.g., Singh and Byerlee 1990).

7. The methodological difficulties of measuring the relationship between irrigation intensity and variability in production are pointed out in Dhawan's (1988:72–73) critique of Hazell's (1982, 1984) finding of a rise from 3.16 to 14.10 percent in the coefficient of variation for food grains in Tamil Nadu state between pre- and post-Green Revolution periods—an increase of 346 percent. By adding some food grains not included in Hazell's calculations, as well as all other crop output, and replacing two drought years during the pre-Green Revolution period that Hazell had removed because "catastrophes of this kind are sufficiently rare and severe . . . that they can be considered as separated phenomena from the more usual year-to-year fluctuations" (Hazell 1982:13), Dhawan calculates coefficients of variation for the earlier (5.56 percent) and later (8.14 percent) periods that show a much smaller increase (1988:72). Contradicting our hypothesis, Dhawan (1988:173) also found that, for most states in India, the coefficient of variation is smaller for irrigated than for nonirrigated production. But, unlike Hazell, Mehra, and others, he concentrates on differences between irrigated and nonirrigated areas for one period of time, and not on the changes accompanying the increase in irrigation and agricultural industrialization through time. Thus, important differences between irrigated and unirrigated areas directly related to irrigation—for example, that irrigation projects are usually sited on the best soils—are not controlled for.

8. Dhawan (1988:154–55) also discusses water source as another important factor in determining variability in the effects of irrigation, based on six years of crop production data from three districts in Tamil Nadu state in India. The coefficient of variation in yield for the district where canal irrigation predominates was 18 percent, while it was 26 percent in the state where hand-dug wells are the most important irrigation water source and 36 percent in the state where tank irrigation is most common.

9. It has been estimated that in the Third World by 1982–1983, 50.7 million hectares were planted in modern varieties of wheat and 72.6 million hectares were planted in modern varieties of rice, or 51.9 and 53.6 percent, respectively, of the total areas planted in those crops (Dalrymple 1986:85–86). Out of the total of 79.1 million hectares planted in maize in the Third World in 1985–1986, 51 percent (40 million hectares) were planted to modern varieties (38 percent in hybrids, and 13 percent in open-pollinated varieties) (Timothy, Harvey, and Dowsell 1988:53–55).

10. In practice, both headender and tailender priority rules are common, as well as

modified and combined versions of these rules (Ostrom 1993). In some systems, for example, a rotation system starts at the head one year, and at the tail the next.

References Cited

Albertson, Maurice L., and Herman Bouwer
  1992   Future of Irrigation in Balanced Third World Development. *Agricultural Water Management* 21:33–44.
Alvares, Claude, and Ramesh Billorey
  1988   *Damming the Narmada: India's Greatest Planned Environmental Disaster.* San Francisco: Third World Network/Asia-Pacific People's Environmental Network.
Anderson, Jock R., and Peter B. R. Hazell
  1989   Introduction. In *Variability in Grain Yields: Implications for Agricultural Research and Policy in Developing Countries*, ed. J. R. Anderson and P. B. R. Hazell, 1–10. Baltimore: Johns Hopkins University Press.
Barker, Randolph, Eric C. Gabler, and Donald Winkelmann
  1981   Long-Term Consequences of Technological Change on Crop Yield Stability: The Case for Cereal Grain. In *Food Security for Developing Countries*, ed. A. Valdés, 53–78. Boulder, Colo.: Westview Press.
Barker, Randolph, and Robert W. Herdt, with Beth Rose
  1985   *The Rice Economy of Asia.* Washington, D.C.: Resources for the Future.
Barlett, Peggy F.
  1989   Industrial Agriculture. In *Economic Anthropology*, ed. S. Plattner, 253–91. Stanford: Stanford University Press.
Batty, J. C., and J. Keller
  1980   Energy Requirements for Irrigation. In *Handbook of Energy Utilization in Agriculture*, ed. D. Pimentel, 35–44. Boca Raton: CRC Press.
Berry, R. A., and W. R. Cline
  1979   *Agrarian Structure and Productivity in Developing Countries.* Baltimore: Johns Hopkins University Press.
Berry, Wendell
  1981   Three Ways of Farming in the Southwest. In *The Gift of Good Land: Further Essays Cultural and Agricultural*, by W. Berry, 47–76. San Francisco: North Point Press.
Brown, Lester R., and John E. Young
  1990   Feeding the World in the Nineties. In *State of the World 1990: A Worldwatch Institute Report on Progress Toward a Sustainable Society*, 59–78. New York: W. W. Norton.
Byerlee, Derek
  1990   Technological Challenges in Asian Agriculture in the 1990s. In *Agricultural Development in the Third World*, 2nd edition, ed. C. K. Eicher and J. M. Staatz, 424–33. Baltimore: Johns Hopkins University Press.
Chambers, Robert
  1980   In Search of a Water Revolution: Questions for Managing Canal Irrigation

in the 1980s. In *Irrigation Water Management,* 23–37. Las Banos, Philippines: International Rice Research Institute.

1984a   Beyond the Green Revolution: A Selective Essay. In *Understanding Green Revolutions: Agrarian Change and Development Planning in South Asia,* ed. T. P. Bayless-Smith and S. Wanmali, 362–79. Cambridge: Cambridge University Press.

1984b   Irrigation Management: Ends, Means and Opportunities. In *Productivity and Equity in Irrigation Systems,* ed. N. Pant, 13–50. New Delhi: Ashish.

1988   *Managing Canal Irrigation: Practical Analysis from South Asia.* Cambridge: Cambridge University Press.

Chibnik, Michael

1990   Double-Edged Risks and Uncertainties: Choices About Rice Loans in the Peruvian Amazon. In *Risk and Uncertainty in Tribal and Peasant Economies,* ed. E. Cashdan, 279–302. Boulder, Colo.: Westview Press.

Cleveland, Cutler J.

1991   Natural Resource Scarcity and Economic Growth Revisited: Economic and Biophysical Perspectives. In *Ecological Economics: The Science and Management of Sustainability,* ed. R. Costanza, 289–317. New York: Columbia University Press.

Cleveland, David A.

1990   Development Alternatives and the African Food Crisis. In *Confronting Change. Stress and Coping in African Food Systems,* Vol. 2, ed. R. Huss-Ashmore and S. Katz, 181–206. New York: Gordon and Breach.

1993   Is Diversity More than the Spice of Life? Diversity, Stability, and Sustainable Agriculture. *Culture and Agriculture* 45–46:2–7.

1995   Diversity, Stability, and Sustainable Agriculture. Unpublished manuscript in possession of author.

1996   Are Large Scale Irrigation Projects Sustainable? The Case of Swabi SCARP, Pakistan. Manuscript under review.

Cleveland, David A., Daniela Soleri, and Steven E. Smith

1994   Do Folk Crop Varieties Have a Role in Sustainable Agriculture? *Bioscience* 44(11)740–51.

Coleman, James

1988   Social Capital in the Creation of Human Capital. *American Journal of Sociology* 91(1):309–35.

Coward, E. Walter, Jr., and Gilbert Levine

1989   Studies of Farmer-managed Irrigation Systems: Ten Years of Cumulative Knowledge and Changing Research Priorities. In *Public Intervention in Farmer-Managed Irrigation Systems,* 1–31. Publication No. 4. Sri Lanka: International Irrigation Management Institute.

Cox, George W., and Michael D. Atkins

1979   *Agricultural Ecology: An Analysis of World Food Production Systems.* San Francisco: W. H. Freeman.

Dahlberg, Kenneth A.

1990   The Industrial Model and Its Impacts on Small Farmers: The Green Revolution as a Case. In *Agroecology and Small Farm Development,* ed. M. A. Altieri and S. B. Hecht, 83–90. Boca Raton: CRC Press.

Dalrymple, Dana G.

1986   *Development and Spread of High-Yielding Wheat Varieties in Developing Countries.* Washington, D.C.: Bureau for Science and Technology, United States Agency for International Development.

Daly, Herman E.

1990   Toward Some Operational Principles of Sustainable Development. *Ecological Economics* 2:1–6.

Daly, Herman E., and John B. Cobb, Jr.

1989   *For the Common Good: Redirecting the Economy Toward Community, the Environment, and a Sustainable Future.* Boston: Beacon Press.

Dhawan, B. D.

1984   Differential Income Impact of Public Canal Irrigation in Maharashtra. In *Productivity and Equity in Irrigation Systems,* ed. N. Pant, 125–45. New Delhi: Ashish.

1988   *Irrigation in India's Agricultural Development: Productivity, Stability, Equity.* New Delhi: Sage Press.

Dover, Michael J.

1985   *A Better Mousetrap: Improving Pest Management for Agriculture.* Washington, D.C.: World Resources Institute.

El Serafy, Salah

1991   The Environment as Capital. In *Ecological Economics: The Science and Management of Sustainability,* ed. R. Costanza, 168–75. New York: Columbia University Press.

Federer, W. T., and B. T. Scully

1993   A Parsimonious Statistical Design and Breeding Procedure for Evaluating and Selecting Desirable Characteristics Over Environments. *Theoretical and Applied Genetics* 86:612–20.

Food and Agriculture Organization

1991   *Production Yearbook 1990,* Vol. 44. Rome: United Nations.

Freeman, David M.

1989   *Local Organizations for Social Development: Concepts and Cases of Irrigation Organization.* Boulder, Colo.: Westview Press.

Gazdar, Muhammad Nasir

1990   *An Assessment of the Kalabagh Dam Project on the River Indus in Pakistan. The Sabz Bagh: "Promising a Rose Garden but Delivering Dust."* Karachi, Pakistan: Environmental Management Society, Pakistan Environmental Network.

Geertz, Clifford

1963   *Agricultural Involution: The Process of Ecological Change in Indonesia.* Berkeley: University of California Press.

Goldschmidt, Walter

1978   *As You Sow: Three Studies in the Social Consequences of Agribusiness.* Montclair, N.J.: Allenheld, Osmun and Company.

Hazell, Peter B. R.
>   1982   *Instability in Indian Foodgrain Production.* Research Report 30. Washington,
>   D.C.: International Food Policy Research Institute (IFPRI).
>   1984   Sources of Increased Instability in Indian and U.S. Cereal Production. *American
>   Journal of Agricultural Economics* 66:302–11.
>   1989   Changing Patterns of Variability in World Cereal Production. In *Variability
>   in Grain Yields: Implications for Agricultural Research and Policy in Develop-
>   ing Countries,* ed. J. R. Anderson and P. B. R. Hazell, 13–34. Baltimore: Johns
>   Hopkins University Press.

Jodha, N. S.
>   1990   *Rural Common Property Resources: A Growing Crisis.* London: International
>   Institute for Environment and Development (IIED).

Lynam, John K., John H. Sanders, and Stephen C. Mason
>   1986   Economic Risk in Multiple Cropping. In *Multiple Cropping Systems,* ed.
>   C. Frances, 250–66. New York: Macmillan.

Mabry, Jonathan B., Hsain Ilahiane, and John R. Welch
>   1991   *Rapid Rural Appraisal of Moroccan Irrigation Systems: Methodological Lessons
>   from the Pre-Sahara.* Report submitted to United States Agency for Interna-
>   tional Development, Morocco, Rabat.

Mandal, M. A. S.
>   1987   Development of Small-Scale Lift Irrigation in Bangladesh. In *Public Intervention
>   in Farmer-Managed Irrigation Systems,* 131–47. Digana, Sri Lanka: International
>   Irrigation Management Institute.

Mehra, Shakuntla
>   1981   *Instability in Indian Agriculture in the Context of the New Technology.* Research
>   Report No. 25. Washington, D.C.: IFPRI.

Merrey, Douglas J.
>   1987   The Local Impact of Centralized Irrigation Control in Pakistan: A Sociocentric
>   Perspective. In *Lands at Risk in the Third World: Local Level Perceptions,* ed.
>   P. D. Little and M. Horowitz, 352–72. Boulder, Colo.: Westview Press.

Netherly, Patricia J.
>   1987   Hydraulics and the Organization of Irrigation in the High Civilizations of the
>   Andean Region of South America. In *Water for the Future: Water Resources De-
>   velopments in Perspective,* ed. W. O. Wunderlich and J. E. Prins, 153–61. Boston:
>   A. A. Balkema.

Netting, Robert McC.
>   1989   Smallholders, Householders, Freeholders: Why the Family Farm Works Well
>   Worldwide. In *The Household Economy: Reconsidering the Domestic Mode of
>   Production,* ed. R. Wilk. Boulder, Colo.: Westview Press.
>   1993   *Smallholder, Householders: Farm Families and the Ecology of Intensive, Sustain-
>   able Agriculture.* Stanford: Stanford University Press.

Norgaard, Richard B., and Richard B. Howarth
>   1991   Sustainability and Discounting the Future. In *Ecological Economics: The Sci-*

ence and Management of Sustainability, ed. R. Costanza, 88–101. New York: Columbia University Press.

Oomen, J. M. V., J. deWolf, and W. R. Jobin

1988   *Health and Irrigation: Incorporation of Disease-Control Measures in Irrigation, a Multi-faceted Task in Design, Construction, and Operation.* Publication No. 45. Wageningen, Netherlands: International Institute for Land Reclamation and Improvement.

Ostrom, Elinor

1992   The Rudiments of a Theory of the Origins, Survival and Performance of Common Property Institutions. In *Making the Commons Work: Theory, Practice, and Policy,* ed. D. W. Bromley, 293–318. San Francisco: ICS Press.

1993   Constituting Social Capital and Collective Action. Paper presented at "Heterogeneity and Collective Action," Workshop in Political Theory and Policy Analysis, Indiana University, Bloomington.

Ostrom, Elinor, Paul Benjamin, and Ganesh Shivakoti

1992   *Institutions, Incentives, and Irrigation in Nepal,* Vol. 1, *Workshop in Political Theory and Policy Analysis.* Bloomington: Indiana University Press.

Pandey, Sushil

1989   Irrigation and Crop Yield Variability: A Review. In *Variability in Grain Yields: Implications for Agricultural Research and Policy in Developing Countries,* ed. J. R. Anderson and P. B. R. Hazell, 234–41. Baltimore: Johns Hopkins University Press.

Park, Thomas K.

1992   Early Trends Toward Class Stratification: Chaos, Common Property, and Flood Recession Agriculture. *American Anthropologist* 94(1):90–117.

Pimentel, David, and Marcia Pimentel

1979   *Food, Energy and Society.* London: Edward Arnold.

Postel, Sandra

1989   *Water for Agriculture: Facing the Limits.* Paper No. 93. Washington, D.C.: Worldwatch Institute.

1992   *Last Oasis: Facing Water Scarcity.* New York: W. W. Norton.

Qasem, Subhi

1990   Issues of Food Security in the Arab Countries. In *Agriculture in the Middle East: Challenges and Possibilities,* ed. A. Salman, 161–77. New York: Paragon House.

Repetto, Robert

1986   *Skimming the Water: Rent-Seeking and the Performance of Public Irrigation Systems.* Paper No. 4. Washington, D.C.: World Resources Institute.

Roberts, Edwin B., Richard H. Cuenca, and Robert M. Hagan

1986   Energy and Water Management with On-Farm Irrigation Systems. In *Energy and Water Management in Western Irrigated Agriculture,* ed. N. K. Whittlesey, 35–71. Boulder, Colo.: Westview Press.

Ryder, Gráinne

1988   China's Three Gorges Project: Whose Dam Business Is It? *Cultural Survival Quarterly* 12(2):17–19.

Ruttan, V. W.

  1984   Induced Innovations and Agricultural Development. In *Agricultural Sustainability in a Changing World Order*, ed. G. K. Douglass, 107–34. Boulder, Colo.: Westview Press.

Sagardoy, J. A.

  1982   *Organization, Operation and Maintenance of Irrigation Schemes*. Irrigation and Drainage Paper 40. Rome: Food and Agriculture Organization of the United Nations.

Saini, G. R.

  1979   *Farm Size, Resource-use Efficiency and Income Distribution: A Study in Indian Agriculture, with Special Reference to Uttar Pradesh and Punjab*. Bombay: Allied Publishers.

Salem-Murdock, Muneera

  1990   Household Production Organization and Differential Access to Resources in Central Tunisia. In *Anthropology and Development in North Africa and the Middle East*, ed. M. Salem-Murdock and M. M. Horowitz, with M. Sella, 95–125. Boulder, Colo.: Westview Press.

Sanders, William T., and Robert S. Santley

  1983   A Tale of Three Cities: Energetics and Urbanization in Pre-Hispanic Central Mexico. In *Prehistoric Settlement Patterns: Essays in Honor of Gordon R. Willey*, ed. E. Vogt and R. Leventhal, 243–91. Albuquerque: University of New Mexico Press.

Scudder, Thayer

  1989   Conservation vs. Development: River Basin Projects in Africa. *Environment* 31(2):4–9, 27–31.

Sheridan, David

  1984   *Cropland or Wasteland: The Problems and Promises of Irrigation*. London: International Institute for Environment and Development.

Singh, A. J., and D. Byerlee

  1990   Relative Variability in Wheat Yields across Countries and Over Time. *Journal of Agricultural Economics* 41:21–32.

Stanhill, G.

  1984   Agricultural Labour: From Energy Source to Sink. In *Energy and Agriculture*, ed. G. Stanhill, 113–30. Berlin: Springer-Verlag.

Strange, Marty

  1988   *Family Farming: A New Economic Vision*. Lincoln: University of Nebraska Press and San Francisco: Institute for Food and Development Policy.

Swearingen, Will D.

  1987   *Moroccan Mirages: Agrarian Dreams and Deceptions, 1912–1986*. Princeton: Princeton University Press.

Tang, Shui Yan

  1992   *Institutions for Collective Action: Self-Governance in Irrigation*. San Francisco: ICS Press.

Timothy, David H., Paul H. Harvey, and Christopher R. Dowsell
  1988   *Development and Spread of Improved Maize Varieties and Hybrids in Develop-
          ing Countries.* Bureau for Science and Technology, United States Agency for
          International Development, Washington, D.C.

Umali, Dina L.
  1993   *Irrigation-Induced Salinity.* Technical Paper No. 215. Washington, D.C.: World
          Bank.

Uphoff, Norman, M. L. Wickramasinghe, and C. M. Wijayaratna
  1990   "Optimum" Participation in Irrigation Management: Issues and Evidence from
          Sri Lanka. *Human Organization* 49(1):26–40.

Wade, Robert
  1986   Common Property Resource Management in South Indian Villages. In *Com-
          mon Property Resource Management,* 231–57. Washington, D.C.: National Re-
          search Council, National Academy of Sciences.
  1988   *Village Republics: Economic Conditions for Collective Action in South India.*
          Cambridge University Press.

Walker, T. S.
  1989   High Yielding Varieties and Variability in Sorghum and Pearl Millet Production
          in India. In *Variability in Grain Yields: Implications for Agricultural Research and
          Policy in Developing Countries,* ed. J. R. Anderson and P. B. R. Hazell, 91–99.
          Baltimore: Johns Hopkins University Press.

Wilk, Richard R.
  1985   Dry Season Agriculture among the Kekchi Maya and Its Implications for Pre-
          history. In *Prehistoric Lowland Maya Environment and Subsistence Economy,* ed.
          M. Pohl, 47–57. Papers of the Peabody Museum of Archaeology and Ethnology,
          Vol. 77. Cambridge: Harvard University.

World Bank
  1992   *World Development Report 1992: Development and the Environment.* Washing-
          ton, D.C.: World Bank and Oxford University Press.

World Resources Institute
  1990   *World Resources, 1990–91.* New York: Oxford University Press.
  1992   *World Resources, 1992–93.* New York: Oxford University Press.

Worster, Donald
  1984   Thinking Like A River. In *Meeting the Expectations of the Land: Essays in Sus-
          tainable Agriculture and Stewardship,* ed. W. Jackson, W. Berry, and B. Colman,
          55–67. San Francisco: North Point Press.
  1985   *Rivers of Empire: Water, Aridity and the Growth of the American West.* New York:
          Pantheon.

Yates, P.
  1981   *Mexico's Agricultural Dilemma.* Tucson: University of Arizona Press.

Contributors

Catherine Besteman is an associate professor of anthropology at Colby College in Waterville, Maine. Her research on Somalia focuses on the people of the Jubba Valley, many of whom are descendants of slaves brought to Somalia in the nineteenth century. Using oral histories, colonial documents, and ethnographic field research, she has traced connections between identity, land tenure, and stratification in the Jubba Valley during the colonial and post-colonial periods. She has published on these topics in the journals *Africa, Ethnohistory,* and the *Journal of Anthropological Research,* and has analyzed the disintegration of the Somali state in *Cultural Anthropology* and *American Ethnologist.* A book co-edited with historian Lee Cassanelli on the patterns of struggles over land in southern Somalia is forthcoming from Westview Press.

Michael E. Bonine is a professor in the Departments of Geography and Near Eastern Studies at the University of Arizona. He conducted fieldwork in Iran from 1969 through 1971 for his Ph.D. dissertation in geography, published as *Yazd and Its Hinterland* (Geographischen Institut der Universitat Marburg, 1980), and did further field research in Iran before the 1978–1979 revolution. His interest in traditional Iranian irrigation and agricultural systems has led to the publication of several articles on qanats, as well as co-editing *Qanat, Kariz and Khattara: Traditional Water Systems in the Middle East and North Africa* (MENAS Press, 1989) with Peter Beaumont and Keith McLachlan. He also is the co-editor, with N. R. Keddie, of *Modern Iran: The Dialectics of Continuity and Change* (State University of New York Press, 1981) and of *The Middle Eastern City and Islamic Urbanism: An Annotated Bibliography of Western Literature* (1994), the latter representing his interest in traditional and modern urbanism in the Middle East. His present research focuses on urban water systems in the Middle East.

David A. Cleveland is an assistant professor in the Department of Anthropology and the Environmental Studies Program at the University of California, Santa Barbara, and co-director of the Center for People, Food and Environment (CFPE) in Tucson. He has worked extensively with Kusasi farmers in northeast Ghana, with Hopi and Zuni farmers in the southwestern United States, and with farmers in northern Pakistan, Egypt, and Mexico. He has been director of the Zuni Sustainable Agriculture Project and the Zuni Folk Varieties Project, and has managed international development projects in the College of Agriculture at the University of Arizona. His current research is on the roles of cultural and biological diversity and stability in agriculture, especially comparison of indigenous and Western approaches to using and developing crop genetic resources. He has authored or co-authored articles on these topics in the journals *Culture and Agriculture, Bioscience,* and

Journal of Ethnobiology. With Daniela Soleri, he co-authored *Food from Dryland Gardens: An Ecological, Nutritional, and Social Approach to Small-Scale Household Food Production* (CFPE, 1991).

Timothy J. Finan is an associate research anthropologist in the Bureau of Applied Research in Anthropology (BARA) at the University of Arizona. His interests lie in agricultural development and economic anthropology, particularly in methods for the systematic study of farming systems, food and livelihood security, the impacts of macro-level policies on rural communities, and participatory development. He has worked extensively in Brazil and Latin America, Lusophone and Francophone Africa, and in the Middle East. He has coauthored several books and monographs, including *Soundings: Rapid and Reliable Research Methods for Practicing Anthropologists* (with John Van Willigen, NAPA Bulletin 10, American Anthropological Association, 1991), and *Portuguese Agriculture in Transition* (with Scott R. Pearson et al., Cornell University Press, 1987). He has also authored and co-authored articles in *American Anthropologist, Human Organization, American Ethnologist, Culture and Agriculture, Journal of Agricultural Economics,* and other journals.

Paul H. Gelles is an assistant professor of anthropology at the University of California at Riverside. Formerly, he held a post-doctoral fellowship at the College of Natural Resources, University of California at Berkeley. He has conducted fieldwork in the central and southern Peruvian Andes, studying, among other things, indigenous narratives, transnationalism, and the way that ethnic identity and cultural politics condition water management. He has published articles on these topics in *American Ethnologist, Allpanchis,* and other journals, and is currently rewriting his dissertation (*Channels of Power, Fields of Contention: The Politics and Ideology of Irrigation in an Andean Peasant Community*) for publication.

Hsain Ilahiane is a doctoral candidate in the Department of Anthropology at the University of Arizona. He has conducted field research on ethnicity and agricultural production in southeastern Morocco, and Native American agriculture in the southwestern United States. His research interests include North African rural development, Saharan ethnohistory and political ecology, arid lands agroecology, oral traditions and historiography, and the anthropology of colonialism. He has served as a consultant for the United States Agency for International Development (USAID), and is a coauthor with Jonathan Mabry and John Welch of *Rapid Rural Appraisal of Moroccan Irrigation Systems: Methodological Lessons from the Pre-Sahara* (USAID-Morocco, 1991).

Mark W. Langworthy is an assistant research scientist in the Department of Agricultural Resource Economics at the University of Arizona. He has conducted field research on development, international trade, and agricultural policy in Africa (Cape Verde, Guinea-Bissau, Mozambique, Chad, and Uganda), Asia (Afghanistan, Bangladesh), and Europe (Portugal). His main research interests include the microeconomics of agricultural households, food security issues, and local community resource management institutions. He has authored and coauthored articles in *Economic Development and Cultural Change,*

American Journal of Agricultural Economics, European Review of Agricultural Economics, and other journals.

J. Stephen Lansing is a professor in the Department of Anthropology and the School of Natural Resources and Environment at the University of Michigan, Ann Arbor. He has carried out fieldwork in Bali and other parts of Indonesia since 1970. Major publications include *Evil in the Morning of the World: Phenomenological Approaches to a Balinese Community* (Michigan Center for Southeast Studies, 1974); *The Three Worlds of Bali* (Praeger, 1983); *Priests and Programmers: Technologies of Power in the Engineered Landscape of Bali* (Princeton University Press, 1991); and *The Balinese* (Harcourt Brace, 1995); as well as several documentary films, including "Three Worlds of Bali" (*Odyssey* 1981) and "The Goddess and the Computer" (Channel 4 London and Documentary Educational Resources, Boston, 1988).

Jonathan B. Mabry, a research archaeologist and project director at the Center for Desert Archaeology in Tucson, has conducted field studies of prehistoric, historic, and contemporary irrigation in the Middle East, North Africa, and the southwestern United States. His research interests include the ecology of early agriculture in arid lands, preindustrial irrigation technologies, and traditional institutions for irrigation management. He has served as a consultant for the United States Agency for International Development (USAID), and is a coauthor of *Rapid Rural Appraisal of Moroccan Irrigation Systems: Methodological Lessons from the Pre-Sahara* (USAID-Morocco, 1991). He has also written several archaeological monographs, and has authored and coauthored articles on his archaeological fieldwork and research in the journals *American Antiquity, Paléorient, Syria,* and *Studies in the History and Archaeology of Jordan.*

Thomas E. Sheridan is the curator of ethnohistory at the Arizona State Museum, University of Arizona. He specializes in the ethnology, ethnohistory, and political ecology of the southwestern United States and northern Mexico. He served as the director of the Mexican Heritage Project at the Arizona Historical Society, where he wrote *Los Tucsonenses: The Mexican Community in Tucson, 1854–1941* (University of Arizona Press, 1986). He has written or edited seven books and monographs, including *Where the Dove Calls: The Political Ecology of a Peasant Corporate Community in Northwestern Mexico* (University of Arizona Press, 1995), and has published articles in the journals *Human Ecology, Journal of the Southwest,* and *Antiquity.* A board member of the Southwestern Mission Research Center (SMRC), he also edits the quarterly *SMRC Newsletter.*

Pamela C. Stanbury is a social science advisor on land and water issues in the Agriculture and Food Security Office of the United States Agency for International Development (USAID). She has conducted field research of traditional irrigation and agricultural development in Sri Lanka, held a post-doctoral fellowship at the International Irrigation Management Institute (IIMI) in Colombo, Sri Lanka, and served as a consultant for USAID. Her research interests include indigenous irrigation institutions, rural social organization. and agricultural development. Her publications concerning irrigation in Sri Lanka include the monograph *Land Settlement Planning for Improved Irrigation Managment: A Case Study of the Kirindi Oya Irrigation and Settlement Project* (IIMI, 1989), and articles in the journal *Research in Economic Anthropology.*

John R. Welch is an archaeologist and historic preservation specialist with the Fort Apache Agency of the U.S. Bureau of Indian Affairs. He is also a doctoral candidate in the Department of Anthropology at the University of Arizona, and for his dissertation is examining the social implications of agricultural dependence at the ancestral Pueblo site of Grasshopper in the eastern Arizona uplands. He has researched the prehistoric agricultural ecology of the Tonto Basin in eastern Arizona, and has conducted ethnographic and archaeological fieldwork in Costa Rica, Belize, Hawaii, and Morocco. He has served as a consultant for the United States Agency for International Development (USAID), and is a coauthor with Jonathan Mabry and Hsain Ilahiane of *Rapid Rural Appraisal of Moroccan Irrigation Systems: Methodological Lessons from the Pre-Sahara* (USAID-Morocco, 1991). He has also authored and coauthored articles in the journals *Kiva* and *American Indian Quarterly.*

Index